7/85 539

D1083464

CONCEPTS OF PARTICLE PHYSICS

Volume I

"Now the smallest Particles of Matter may cohere by the strongest Attractions and compose bigger Particles of weaker Virtue; and many of these may cohere and compose bigger Particles whose Virtue is still weaker, and so on for diverse successions, until the Progression ends in the biggest Particles on which the Operations in Chymistry and the Colours of natural Bodies depend, and which by cohering compose Bodies of a sensible Magnitude.

There are therefore Agents in Nature able to make the Particles of Bodies stick together by very strong Attractions. And it is the Business of Experimental Philosophy to find them out."

Isaac Newton, *Opticks,* 1704

CONCEPTS
OF
PARTICLE PHYSICS

Volume I

KURT GOTTFRIED

Cornell University

VICTOR F. WEISSKOPF

Massachusetts Institute of Technology

CLARENDON PRESS · OXFORD

OXFORD UNIVERSITY PRESS · NEW YORK

1984

Oxford University Press, Walton Street, Oxford OX2 6DP

London   Glasgow   New York   Toronto
Delhi   Bombay   Calcutta   Madras   Karachi
Kuala Lumpur   Singapore   Hong Kong   Toyko
Nairobi   Dar Es Salaam   Cape Town
Melbourne   Wellington

and associate companies in
Beirut   Berlin   Ibadan   Mexico City

Published in the United States by
Oxford University Press, Inc., New York

Library of Congress Cataloging in Publication Data

Gottfried, Kurt, 1929–
Concepts of particle physics.

Bibliography: v. 1, p.
Includes index.
1. Particles (Nuclear physics)   I. Weisskopf, Victor
Frederick, 1908–   II. Title.
QC793.2.G68   1984   539.7'21   83-17275
ISBN 0-19-503392-2 (v. 1)

Printing (last digit): 9 8 7 6 5 4 3

Printed in the United States of America

To the men and women who create the accelerators,
the detectors, and the experiments from which
the concepts of particle physics spring.

PREFACE

> . . . Call me a fool;
> Trust not my reading nor my observation,
> Which with experimental seal doth warrant
> The tenour of my book.
>
> *Much Ado about Nothing,* Act IV

The goal of this book is to elucidate basic and well-established concepts of particle physics to those who do not have the sophisticated training in mathematics and physics that is habitually expected of students of this subject. For that reason this book does not really belong to what we may call the written tradition of physics.

Like every intellectual pursuit, physics has both a written and an oral tradition. Intuitive modes of thought, inference by analogy, and other strategems that are used in the effort to confront the unknown are transmitted from one generation of practitioners to the next by word of mouth. After the work of creation is over, the results are recorded for posterity in a logically impeccable form, but in a language that is often opaque. The beginner is expected to absorb this written tradition, and only the survivors of this trial-by-ordeal are admitted to circles where the oral tradition is current. We could only hope to strive toward our goal by leaning heavily on the oral tradition. That is not meant as an apology, for we believe that this tradition plays an essential role not only in the creation of physics, but also in the search for a deeper understanding. A wider dissemination of the oral tradition is therefore in order.

What we expect from our readers can perhaps be best conveyed by saying that this book had its origin in a lecture series that we have given over a period of many years to students who have won scholarships to spend a summer at CERN. These students had completed their first degree in a variety of scientific and engineering disciplines at universities throughout Europe. We were therefore able to assume a knowledge of electrodynamics, relativity, and elementary quantum mechanics at a level comparable to what is now standard at the senior level in American undergraduate curricula. That is the background expected from readers of Volume I of this book. Volume II only has Volume I as a formal prerequisite; nevertheless, it is more demanding and sophisticated than Volume I.

Elementary quantum mechanics is definitely a prerequisite for the understanding of this book. Since our readers are likely to have a rather heterogeneous preparation in this subject, we have, in Part B of Volume I, provided a terse summary of the principles of quantum mechanics in a form adapted to the needs of particle physics.

vii

Having explained what we expect from our readers, we now describe what our readers should and should not expect from us.

First, they should recognize that this is not a textbook in the conventional sense. There are no homework exercises, and very few detailed calculations of elementary examples, complete with factors of 2π, etc. On the whole, our attention is focused on the general conceptual framework. This approach is buttressed by order-of-magnitude estimates, symmetry considerations, selection rules, and other arguments that permit one to understand the characteristic behavior of physically important quantities that emerge from full-fledged, "professional" calculations.

Second, while we provide a reasonably detailed overview of the experimental data in the figures, in Chapter III of Volume II, and in Appendix I, we say hardly anything about the prodigiously sophisticated experimental techniques that have yielded this rich bounty of facts. Readers who wish to gain some acquaintance with these techniques should consult the excellent text by Perkins (1982).

Third, within the constraints already explained, we try to paint a fairly accurate, and therefore rather conservative, picture of the current status of the field. In so doing, we have taken into account the remarkable growth of knowledge during the past decade—a virtual revolution that has led us to write a book that bears but little resemblance to what we envisaged when we began work on the manuscript in the summer of 1974. On the other hand, we have shied away from discussing the large number of intriguing speculations that this revolution has spawned. The only exception to this rule is Section 13 of Volume I, which sketches those ideas that appear to be most fruitful and provocative at this time.

To summarize, this book is primarily intended for the autodidact who is curious about recent developments in fundamental physics, whether he or she be a student of any branch of physics or a professional scientist in a discipline other than particle physics. When complemented with suitable monographs, the book should also be useful to students who wish to pursue research in particle physics, and we even entertain the hope that our colleagues in the field itself will, on occasion, find here new ways of seeing phenomena that they understand and live with daily.

The book has been divided into two volumes. Volume I is a self-contained overview of the whole subject, beginning with post-Renaissance concepts, and ending with speculations concerning the relationship between particle physics and cosmology. Volume II goes over much of the same material again, but at a deeper level. It can be traversed by a variety of multiply connected paths that all begin from an assumed knowledge of Volume I, and which are described in the Preface to Volume II.

Our primary objective throughout has been pedagogy, not history. For that reason we have not given references to the original literature, nor have we tried to face the exceedingly difficult task of assigning credit for the theoretical and experimental work that is described, for that would have taken, at an absolute minimum, a plethora of enormous footnotes which would not be of interest to the vast majority of our readers. One should therefore recognize that the experimental data shown in the figures and tables are, with but few exceptions, not the results of the experiments that discovered the phenomena in question. In the same vein, references to theoretical papers are not to the original breakthrough contributions, but to pedagogic articles, or to recent comparisons between theory and experiment. We hope that our colleagues in the field will not take this amiss. The most distinguished amongst them should take solace from the fact that it is no longer customary to refer to a publication by Galileo when one exploits the momentum conservation law.

Many of our colleagues on both sides of the Atlantic have provided us with invaluable insights and criticism. We are especially indebted to Valentine Telegdi who, for a time, was to be our co-author. Others who have given generously of their knowledge are John Bell, Sidney Coleman, Sidney Drell, Richard Feynman, Erwin Gabathuler, Vladimir Glaser, Alan Guth, Maurice Jacob, Robert Jaffe, Kenneth Johnson, Toichiro Kinoshita, Gething Lewis, Michael Peskin, Carlo Rubbia, Alvaro de Rujula, Heinrich Wahl, Klaus Winter, Tung-Mow Yan, and Donald Yennie. We are grateful to the Particle Data Group at Berkeley for providing us with an early version of their latest compilations and for other valuable data. William Lock gave us helpful advice concerning publication of this book. We owe a special note of thanks to Erich Rathske, for his meticulous help with proofreading has made this a much more accurate book than it would otherwise have been. The manuscript has gone through so many versions that we cannot express our appreciation to all who have helped in its preparation, but we are particularly indebted to Diane Eulian, Donald Miller, Jenni Morris, Velma Ray, and Milda Richardson.

The generous support of CERN throughout the years is gratefully acknowledged.

December 1983
Ithaca, New York   K.G.
Cambridge, Massachusetts   V.F.W.

NOTATION AND CITATIONS

This book is divided into Chapters, denoted by roman numerals (I, II, etc.). Chapter I constitutes Volume I. Chapters are divided into Parts, designated by A, B, etc., and Parts are subdivided into Sections and Subsections, enumerated as 1(a), 1(b), ... , 2(a), 2(b), ... , etc. The enumeration of equations begins afresh in every Part of each Chapter. A purely numerical reference to an equation, as in Eq. (41), or a Section, as in §4(a), refers to an equation or Section in the *same* Part. When an equation or Section in *another* Part (say C) of the *same* Chapter is referred to, the citation would read Eq. C(41) or §C.4(a), while if the reference is to *another* Chapter (say IV), the citations would read Eq. IV.C(41) or §IV.C.4(a).

Figures are numbered afresh in each chapter and are referred to by that number within each Chapter. A figure (e.g., No. 3) in another Chapter, say II, is denoted by Fig. II.3.

References to the Bibliography (there is one at the end of each volume) are cited using first author and year of publication, as in (Dirac, 1958). Unless there is a specific reference, experimental data are taken from the Particle Data Group, 1982; Appendix I is an abbreviated version of this data compilation.

Text in small type, which is always set between bracket symbols, [[ ]], is more advanced than the surrounding material, or only of secondary importance at that juncture, and can be skipped at a first reading.

Particle reactions are usually written in an abbreviated form, as in $\gamma A \to A e^+ e^-$, which means $\gamma + A \to A + e^+ + e^-$, where A stands for an atomic nucleus.

With but rare exceptions, we use *natural units*, wherein $\hbar = c = 1$; this system of units is explained in I.B.1(e). Other commonly used units are fm $= 10^{-13}$ cm, Å $= 10^{-8}$ cm, MeV $= 10^6$ eV, and GeV $= 10^9$ eV.

Our quantum mechanical notation is defined in §I.B.1. We designate everyday Euclidean 3-space by $\mathscr{E}_3$, while *abstract* Euclidean 3-spaces carry a superscript, such as T, which stands for weak isospin, or I, which stands for hadronic isospin. A complex N-dimensional vector space is designated by $\mathscr{C}_N$. The notation $\{...\}$ refers to the set of objects ....

Vectors in $\mathscr{E}_3$ are in roman boldface: **E, p, ϵ,** etc.; unit vectors have a caret, as in $\hat{\boldsymbol{\epsilon}}$. Minkowski 4-vectors are denoted by italic boldface, as in $\boldsymbol{x} = (t, \mathbf{r})$ or $\boldsymbol{p} = (E, \mathbf{p})$. The Lorentz-invariant scalar product of two 4-vectors is written as a dot product, as in $\boldsymbol{x} \cdot \boldsymbol{p} = Et - \mathbf{r} \cdot \mathbf{p}$. Vectors in an *abstract* $\mathscr{E}_3$ space are written as $\vec{A}, \vec{B}$, etc., and their scalar products $\vec{A} \cdot \vec{B}$, etc., have the usual meaning. Beginning with Chapter IV, we also use the notation $\vec{A}, \vec{B}$, etc., to represent color-$SU(3)$ octets.

CONTENTS

I

BASIC CONCEPTS

A. THE EVOLUTION OF THE PARTICLE CONCEPT BEFORE THE ADVENT OF QUANTUM MECHANICS

The idea that matter consists of some simple and unchanging elementary constituents is deeply ingrained in our way of thinking. We observe that matter appears in an enormous variety of different realizations, qualities, shapes, and forms, transforming from one into others. In these changes, however, we observe many recurring properties—many features that remain unchanged, or, if changed, that recur under similar conditions. We find constancies and regularities in the flow of events; we recognize materials with well-defined properties, such as water, metals, rocks, or living species; we conjecture that there must be something unchanging in nature that causes these recurrent phenomena. This is the origin of the idea of elementary particles, as Newton expressed it so lucidly:

> All these Things being considered, it seems probable to me that God in the beginning formed Matter in solid, massy, hard, impenetrable, moveable Particles, of such Sizes and Figures, and with other Properties, and in such Proportion to Space, as most conduced to the End for which He formed them; and that these primitive Particles being Solids are incomparably harder than any porous Bodies compounded of them; even so very hard, as never to wear or break in Pieces; no ordinary Power being able to divide what God himself made in the first Creation.... And therefore that Nature may be lasting, the Changes of corporeal Things are to be placed only in the various Separations and new Associations and Motions of these permanent Particles.

Newton put forward a seminal assumption concerning elementary particles: they must have well-defined, specific, unchanging properties. In his time, this quality could only result from being "incomparably" hard. Such elementary units were eventually discovered when chemists during the 18th and 19th centuries found that all matter is made up of 92 different species of "atoms," a term that is the Greek equivalent* of Newton's "incomparable hardness."

* Atomai = uncuttable.

3

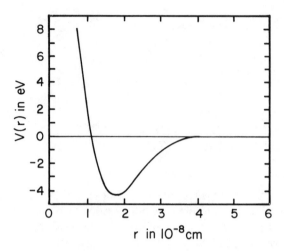

Fig. 1. Rough sketch of the potential of the chemical force between two atoms. The quantitative details depend on the kind of atoms. Their separation r is in units of ångströms, 10^{-8} cm.

In order to describe the properties of matter as we observe them in our terrestrial environment, the chemists and physicists of the 19th century postulated several distinct kinds of forces acting between atoms. Here is a partial list of those forces:

1. Chemical forces between atoms, which keep them together so that they form specific groups of atoms in the form of molecules or solids. These forces are attractive over a range of a few ångströms ($1 \text{ Å} = 10^{-8}$ cm), but fall off exponentially outside that range. At distances shorter than that range, they become repulsive and prevent the atoms from overlapping one another. A rough picture of the potential energy of these forces is given in Fig. 1.
2. Forces acting between molecules which
   - keep the molecules bound to each other in solids and liquids;
   - act repulsively when molecules get very near to each other and are responsible for molecular collisions in gases, and the resistance of solids and liquids against compression;
   - provide a weak attraction between molecules, such as hydrogen bonds and the Van der Waals forces;
   - produce capillary forces on the surface of liquids;
   - result in adhesion between solid surfaces.
3. Electrical forces between charged atoms and molecules (ions).
4. Gravitational attractions between atoms or molecules, which are negligible compared to all the other forces as long as the body in question is not macroscopic.

The concepts of atoms and molecules made it possible to explain the phenomena of heat and temperature as being due to atomic or molecular random motion. The successes of this idea and, in particular, the quantitative explanation of the behavior of dilute gases, gave great support to the hypothesis that atoms and molecules really exist.

Toward the end of the 19th century, profound problems and difficulties appeared within the framework of these ideas. The ground was prepared for the revolution that culminated in the creation of quantum mechanics. These difficulties can be divided into internal and external ones, the former being logical inconsistencies within the theories existing at that time, the latter due to experimental results which seemed to contradict the theories.

The most important of these internal problems was what we shall call the *Boltzmann paradox*. Boltzmann was able to prove that when a system is in thermal equilibrium, each and every one of its degrees of freedom will contribute the amount $\frac{1}{2}kT$ to the total energy, where T is the absolute temperature and k is the Boltzmann constant $[k \simeq 8.6 \times 10^{-11} \text{ MeV } °\text{K}^{-1}]$. This is the well-known *equipartition theorem*. The specific heat of an aggregate should therefore be independent of temperature, and equal to $\frac{1}{2}nk$, where n is the number of degrees of freedom of this system per unit mass— the number of independent ways in which it can move. For example, an ideal gas of n diatomic molecules per unit mass has $5n$ degrees of freedom if one assumes that the molecules are rigid, because each molecule can move as a whole in three directions, while its orientation is given by an axis whose direction is determined by two angular coordinates. This result was verified at ordinary temperatures. But at very low temperatures the specific heat decreased, in contradiction with the theory. What was worse, however, was that the basic assumption of only five degrees of freedom per diatomic molecule proved to be questionable. Why is there no internal motion within the molecule, such as vibrations of the atoms, or deformations of the molecule?

This is just one example of the contradiction in principle which we have called the Boltzmann paradox. In classical physics any piece of matter of any size—even an atom—must have an undetermined number of degrees of freedom: it must be able to undergo deformations of all kinds. The equipartition theorem applies to any degree of freedom, even those that describe very stiff deformations; they too would contribute $\frac{1}{2}kT$ to the specific heat. Hence, within classical kinetic theory, the specific heat is undetermined, because the number of degrees of freedom of any finite chunk of continuous matter is infinite, whatever its size.

The same paradox appears even more forcefully in the thermodynamics of the classical radiation field. The number of degrees of freedom of the radiation field in a given volume of empty space is infinite, because an infinite span of frequencies is available. Since every material body at finite temperature emits some radiation, it follows from the theory that it is impossible to establish thermal equilibrium between such a body and its own radiation if the body is placed in a perfectly reflecting enclosure. Energy would flow into the insatiable radiation field with its infinite number of degrees of freedom. The classical equipartition theorem therefore destroys the theory on which it is based.

We now come to the external difficulties arising from experimental results that contradicted the accepted theories. One was mentioned already: the specific heat of gases and solids decreased markedly at low temperatures (usually below 100° Kelvin), in contradiction with the equipartition theorem, as if matter does not like to absorb small amounts of heat energy.

Other difficulties arose from the study of electricity and optics. At the end of the 19th century the electron was discovered, identified as the carrier of currents in metals, and as an important constituent of atoms. Electrons within atoms were recognized as being responsible for the radiation phenomena involving matter. In particular, the splitting of spectral lines in a magnetic field (the Zeeman effect) made it plausible that radiant matter produces light by vibrating electrons. The existence of spectral lines proved that light is emitted by matter with well-defined frequencies which are characteristic of the material. It was therefore natural to assume that electrons are elastically bound in atoms and perform oscillatory vibrations with frequencies that are characteristic of each material. But again the equipartition theorem caused trouble, because it predicts radiative properties of matter in glaring contradiction with the facts. The theorem requires that the electrons should oscillate with *all* their characteristic frequencies at *any* temperature, but with amplitudes that decrease continuously toward zero as the temperature is lowered. In fact, however, when heated, a piece of matter glows red, then yellow, and finally bluish-white: Ever higher frequencies are emitted as the temperature is raised. It was not generally appreciated that the classical theories contradict the observations of every blacksmith and cook. As so often in the history of science, the conflict between simple and generally known facts, and current theoretical ideas, was recognized only slowly. It took many decades, and a continuous barrage of new facts, before it became clear that the nature of the atom cannot be explained by the concepts of classical physics.

The most elementary experiments of chemistry are also at variance with classical principles: They are interpreted in terms of atoms and molecules exhibiting properties characteristic of the species, all members of one kind being completely identical with each other. This identity of properties and behavior survives collisions in gases, tight packing in liquids or solids, and subsequent evaporation; it can be recovered after spark discharges, ionization, and all other violent disturbances. When the original conditions are restituted, the atoms or molecules regenerate themselves with identical properties, without any trace of the previous history. These features are utterly alien to the ideas of classical physics, where initial and boundary conditions are essential ingredients in determining the motion and structure of every system.

Any classical model of the atom had to assume the existence of nonelectrical forces that kept the electrons in fixed positions when the atom was undisturbed, because Earnshaw's theorem proves that no system of charges interacting only via Coulomb's law can be in static equilibrium, while any moving system of charges radiates and thereby loses energy unendingly. The assumption of nonelectric forces was not disturbing in itself, because, as we have seen, all sorts of forces had already been invoked for a host of purposes. The fatal blow against such views of atomic structure and dynamics came as late as 1911 with the discovery of large-angle scattering of α-particles by metallic foils. This was immediately interpreted as proving that most of the mass and positive charge of the atom is concentrated in a small nucleus. The notion that electrons are elastically bound by a conspiracy of electrical and other forces was seen to be highly implausible; rather, it became natural to suppose that the electrons revolve about the nucleus like planets around the sun. But this picture also showed that there was no hope of understanding the atom within the framework of classical mechanics and electrodynamics. A planetary atom would not emit light with selected definite frequencies. Furthermore, a classical planetary atom does not have a well-defined structure that is stable against collisions, whereas atoms manifestly have that property. What is perhaps worse, such a system could not exist for any reasonable length of time: Within 10^{-9} sec electrons in orbits of atomic dimensions (~ 1 Å) would spiral into the nucleus because of energy loss through radiation, producing an inert and electrically neutral body consisting of electrons sitting on the nuclear surface.

The discovery of the nucleus quickly led to the Bohr–Rutherford model of the atom. This highly successful model had two essential ingredients: a purely electromagnetic interaction between all atomic constituents, and the quantization of angular momentum, which also implied the quantization of energy. The quantum concept had been developed during the preceding decade, principally in connection with the emission and absorption of light. This was its first use in what is, in essence, a problem of mechanics, not of electrodynamics. Quantization of energy had one immediate beneficial consequence: it resolved the Boltzmann paradox.

The Bohr–Rutherford model, like all facets of the "old quantum theory" which foreshadowed quantum mechanics, was based on a set of ad hoc rules that constituted clear violations of the basic principles of classical mechanics. The partial success of the old quantum theory showed that atomic physics could not be understood within the classical framework—that that framework would have to be rebuilt from its very foundations.

The discovery of radioactivity at the end of the 19th century shook physicists even more than the complex web of observations concerning the chemical, electrical, and optical properties of the atom. Here they observed particles and radiations with energies that exceeded the energies they had associated with processes in the atom by factors of the order of a million. The law of the conservation of energy was now in doubt; there was a heightened feeling of mystery surrounding the structure of atoms. It was not yet realized that these phenomena point to the next level of insight: the internal structure of the nucleus. Despite that, radioactivity provided powerful tools which were necessary for penetrating into the structure of atoms. The existence of atomic nuclei cannot be inferred from optics or chemistry; the discovery of the nucleus was contingent on the earlier discovery of radioactive α-emitters.

B. NONRELATIVISTIC QUANTUM MECHANICS
AND ATOMIC PHYSICS

During a few hectic years, from 1925 to 1927, the hodgepodge of recipes and rules that constituted the old quantum theory was replaced by a new mechanics—quantum mechanics. It has the logical consistency and completeness that we expect of a fundamental theory of physics. While readers of this book are expected to have a reasonable knowledge of nonrelativistic quantum mechanics, we shall provide a résumé so as to establish a common language. We shall assume no knowledge of relativistic quantum mechanics or of quantum field theory. Indeed, it is one of the major purposes of this book to explain the basic principles of that subject, for it provides many of the key concepts of particle physics.

1. The principles of quantum mechanics

Consider some physical system $\mathcal{S}$. For the sake of concreteness, we shall, on occasion, think of the hydrogen atom, an electron in the field of a proton.

(a) State vectors. Observables

Every possible state of $\mathcal{S}$ is described by a *state vector* $|\psi\rangle$. To every $|\psi\rangle$ we ascribe an adjoint vector $\langle\psi|$. If $|\psi_1\rangle$ and $|\psi_2\rangle$ are two state vectors, their scalar product $\langle\psi_2|\psi_1\rangle$ is a complex number satisfying $\langle\psi_2|\psi_1\rangle^* = \langle\psi_1|\psi_2\rangle$. The quantity $P_{21} = |\langle\psi_2|\psi_1\rangle|^2$ is the *probability* that $\mathcal{S}$ will be observed to be in the state $|\psi_2\rangle$ when it is prepared in the state $|\psi_1\rangle$. For that reason the scalar product is called the *probability amplitude*. Obviously $0 \leqslant P_{21} \leqslant 1$, and $|\langle\psi|\psi\rangle| = 1$.

If a and b are complex numbers satisfying $|a|^2 + |b|^2 = 1$, $a|\psi_1\rangle + b|\psi_2\rangle$ is again a state vector—it too describes a possible state of $\mathcal{S}$. The totality of all state vectors form the Hilbert space $\mathfrak{H}$.

To any *observable* quantity associated with $\mathcal{S}$ (e.g., angular momentum or position of the electron) there corresponds an *operator* A. When acting on $|\psi_1\rangle$, A produces another vector $A|\psi_1\rangle$, and its scalar product with $|\psi_2\rangle$, $\langle\psi_2|A|\psi_1\rangle$, is called a *matrix element* of A. The vector adjoint to $A|\psi_1\rangle$ is the vector $\langle\psi_1|A^\dagger$, where the Hermitian adjoint, $A^\dagger$, is defined by

$$\langle\psi_2|A|\psi_1\rangle^* = \langle\psi_1|A^\dagger|\psi_2\rangle.$$

The diagonal matrix element $\langle\psi|A|\psi\rangle$ is of special importance: it gives the *expectation value* of A in the state $|\psi\rangle$. By this one means that measurements of the observable A on an ensemble of system $\mathscr{S}$, all identically prepared to be in the state $|\psi\rangle$, will yield $\langle\psi|A|\psi\rangle$ as the statistical mean of these measurements. Since the expectation value must be real in any state, observables are represented by operators satisying $A = A^\dagger$. Such operators are said to be Hermitian.

We shall frequently encounter *transition probabilities* and *transition amplitudes*. These concepts are most readily understood with the help of an example. Let $|\psi\rangle$ represent an electron of momentum $\mathbf{p}$ incident on a nucleus, together with the scattered waves resulting from the Coulomb interaction. The probability that electrons prepared in $|\psi\rangle$ are to be found in a plane wave state $|\chi\rangle$ of definite momentum $\mathbf{p}'$ is called the probability for the transition from $\mathbf{p}$ to $\mathbf{p}'$. It is given by $|\langle\chi|\psi\rangle|^2$, and $\langle\chi|\psi\rangle$ is called the transition amplitude.

(b) The equations of motion

Of all the observables belonging to $\mathscr{S}$, one is especially significant, because it governs the time development of the system. This is the operator H, the *Hamiltonian*, and it determines that development by means of the *Schrödinger equation*:

$$i\hbar \frac{\partial}{\partial t}|\psi(t)\rangle = H|\psi(t)\rangle. \tag{1}$$

This has the formal solution

$$|\psi(t)\rangle = e^{-iH(t-t_0)/\hbar}|\psi(t_0)\rangle, \tag{2}$$

where $|\psi(t_0)\rangle$ is the state at some initial time t_0. Henceforth we set $t_0 = 0$.

The expectation value of any observable A in the evolving state is then given by

$$a(t) = \langle\psi(t)|A|\psi(t)\rangle \tag{3}$$

$$= \langle\psi(0)|e^{iHt/\hbar}Ae^{-iHt/\hbar}|\psi(0)\rangle. \tag{4}$$

These equations show that we can use two equivalent though different "pictures" for describing time evolution. In the *Schrödinger picture*, the observables A are fixed, and the state vectors $|\psi(t)\rangle$ revolve in Hilbert space. In the *Heisenberg picture*, the state vectors are fixed at, say, $t_0 = 0$, but the observables move:

$$A(t) = e^{iHt/\hbar}Ae^{-iHt/\hbar}. \tag{5}$$

Obviously

$$i\hbar \frac{\partial}{\partial t}A(t) = [A(t),H], \tag{6}$$

where $[A,B] = AB - BA$ is the commutator. Equation (6) plays the role of the equation of motion in the Heisenberg picture. The Schrödinger picture is the more familiar one, but the Heisenberg picture is more appropriate in quantum field theory, where the space and time specifications of events are on an equal footing.

(c) Unitary operators and symmetries

Unitary operators are of great importance in quantum mechanics. An operator U is said to be unitary if $UU^\dagger = U^\dagger U = 1$, or $U^\dagger = U^{-1}$. Clearly the operator e^{iA} is unitary if A is Hermitian. The time evolution of (2) is given by the unitary operator

$$U(t) = e^{-iHt/\hbar}. \tag{7}$$

Consider some symmetry operation, such as a spatial translation. Let $|\psi_{i,j}\rangle$ be two different states of $\mathcal{S}$, for example, two electron wave packets running in different directions through the same region, so that $P_{ij} = |\langle\psi_i|\psi_j\rangle|^2$ does not vanish. If $|\psi_i\rangle$ and $|\psi_j\rangle$ are precisely the same states as seen in a coordinate frame that has been translated with respect to the first, the probability $|\langle\psi_i'|\psi_j'\rangle|^2$ must still be P_{ij}, and by an appropriate choice of phases, we can therefore make $\langle\psi_i|\psi_j\rangle = \langle\psi_i'|\psi_j'\rangle$. This demonstrates that the translation that relates $|\psi_i\rangle$ to $|\psi_i'\rangle$ is given by a unitary operator W, $|\psi_i'\rangle = W|\psi_i\rangle$, because $\langle\psi_i'|\psi_j'\rangle = \langle\psi_i|W^\dagger W|\psi_j\rangle = \langle\psi_i|\psi_j\rangle$. Hence *symmetry transformations are represented by unitary operators.*[*]

(d) Stationary states

The Schrödinger equation possesses solutions where the time development (2) is given by a numerical phase factor instead of the unitary operator $e^{-iHt/\hbar}$:

$$|\psi(t)\rangle = e^{-iEt/\hbar}|\psi_E\rangle. \tag{8}$$

The number E and *the stationary state* $|\psi_E\rangle$ are determined from the time-independent Schrödinger equation

$$(H - E)|\psi_E\rangle = 0. \tag{9}$$

Thus E is an eigenvalue of the operator H, and $|\psi_E\rangle$ the corresponding eigenvector.

The other observables also have *eigenstates* and *eigenvalues*. For the generic operator A we write $(A - a)|a\rangle = 0$, where a is the eigenvalue and $|a\rangle$ the eigenstate. In general $|a\rangle$ and $|\psi_E\rangle$ will not be identical. Eigenvalues are often called *quantum numbers*.

---

[*] There is one exception to this statement: time reversal is not described by a unitary operator. For a discussion of this point, see Gottfried (1966), §§27 and 39, and Wick (1966).

If $\mathscr{S}$ has more than one degree of freedom, the energy will not suffice to determine a unique state vector: there are many—often an infinite number—of states belonging to an eigenvalue E. Quite generally, if $\mathscr{S}$ has n degrees of freedom when viewed as a classical system, n distinct eigenvalues are required to specify a state vector uniquely. Thus if $\mathscr{S}$ is our electron, and we ignore spin for the moment, $n = 3$, and the eigenvalues of three distinct observables are required. These could be the energy E, the orbital angular momentum l, and the projection m of the angular momentum along some axis. This fully specified stationary state could be designated by $|Elm\rangle$. Another completely specified state is $|\mathbf{r}\rangle$, where $\mathbf{r} = (x,y,z)$ are the three eigenvalues of the operators for the electron's cartesian coordinates. But $|\mathbf{r}\rangle$ is not a stationary state: it describes an electron that is at the point $\mathbf{r}$ with certainty. The Schrödinger *wave function* is the probability amplitude

$$\psi_{Elm}(\mathbf{r}) = \langle \mathbf{r}|Elm\rangle; \tag{10}$$

$|\psi_{Elm}(\mathbf{r})|^2$ has the familiar probability interpretation.

The totality $\{a\}$ of all eigenvalues of an observable A is called the *spectrum* of A. In particular, $\{E\}$ is the energy spectrum. A spectrum may be discrete, continuous, or both. When $\mathscr{S}$ is the electron in a Coulomb field, the energy spectrum is continuous for $E > 0$, and discrete for $E < 0$. The stationary continuum states describe scattering in a Coulomb field. The discrete states are the bound hydrogenic states.

The continuous states have an infinite degeneracy, because the energy of a scattering state does not depend on the orientation of the incoming momentum, only on its magnitude. The discrete (bound) states only have a finite degeneracy, however. The lowest lying bound state is called the ground state.

In constructing a complete set of n observables to describe states uniquely, proper attention must be paid to the *uncertainty principle*. If A and B are two observables that do not commute, their uncertainty product $\Delta A \Delta B$ does not vanish, and they cannot be specified simultaneously. Consequently the set of observables used to specify states must be a commuting set.

The stationary states form a complete set. By that one means the following: Let $\{|Ea\rangle\}$ be the stationary states, where a stands for all the eigenvalues apart from energy that are needed to specify the state vector uniquely; then *any* state can be written as some linear superposition of the stationary states.

While the stationary states—and eigenstates quite generally—are crucially important, it should not be forgotten that such states cannot describe all physical situations. Nevertheless, a nonstationary state can be written as a linear superposition of stationary states, or, for that matter, in terms of any complete set.

An important class of states occurs in collision theory. The evolving state vector $|\psi(t)\rangle$ describes a projectile in a localized wave packet initially running toward the target, and after the collision, outgoing scattered waves. If the wave packet has transverse dimensions large compared to the target, and a well defined momentum **p**, the collision probability can be computed with a stationary state, $|\psi_\mathbf{p}\rangle$. The latter belongs to the continuous spectrum of H. Its wave function $\langle \mathbf{r}|\psi_\mathbf{p}\rangle$ is a superposition of a plane wave $\exp(i\mathbf{p}\cdot\mathbf{r})$ describing the incoming projectile, and an outgoing spherical wave.

(e) The natural system of units

In particle physics it is convenient to use the total energy, including the mass energies of the constituents, instead of an energy measured with respect to that of the separated constituents, as is customary in atomic physics. It is also more convenient to introduce a special system of units. Instead of a system based on mass, length, and time, such as the cgs-system, we choose a system based on mass, action, and velocity. The units of the latter two quantities will be Planck's constant $\hbar$, and the speed of light c; the unit of mass is left open. Since $\hbar$ and c appear to be fundamental constants, it is natural to express every quantity in these units. Here is a list of a few important quantities expressed in this way:

$$
\begin{array}{llll}
\textit{energies} & \text{in units of } c^2 & E = E'c^2 & \\
\textit{momenta} & \text{in units of } c & p = p'c & \\
\textit{lengths} & \text{in units of } \hbar/c & l = l'\hbar/c & \\
\textit{times} & \text{in units of } \hbar/c^2 & t = t'\hbar/c^2 & (11) \\
\textit{velocities} & \text{in units of } c & v = v'c & \\
\textit{charges} & \text{in units of } (\hbar c)^{\frac{1}{2}} & e^2 = e'^2\hbar c & \\
\textit{angular momentum} & \text{in units of } \hbar & J = J'\hbar & \\
\end{array}
$$

All the primed quantities are either dimensionless, or some power of a mass. Energy and momentum (E' and p') have the dimension of a mass; length and time (l' and t') have the dimension (mass)$^{-1}$; and velocity, charge, and angular momentum (v',e',J') are dimensionless. The electronic charge e' has the value $(4\pi/137.04)^{\frac{1}{2}}$ in these units.* From now on we omit the prime and use the new units in most instances.

When using these units, we are still free to choose the unit of mass with which we desire to express quantities. This unit is conveniently chosen as the mass of a particle that is important in the problem, such as the proton, the electron, or the pion.

2. Resolution of the Boltzmann paradox

Quite generally, any system of interacting particles which is able to form a bound state—a state with less energy than that of its separated constituents at rest—can do so only with certain discrete values of energy. The states of such

---

* We use "rationalized" units, because we shall have more to do with Maxwell's equations than with Coulomb's law. In rationalized units, the Coulomb potential is $e^2/4\pi r$, but Maxwell's equations are free of 4π's; the energy density of the field is $\frac{1}{2}(\mathbf{E}^2 + \mathbf{B}^2)$.

systems have well-defined properties, which depend on the nature of the particles and their interactions.

A system in its ground state can be considered to be endowed with fixed, unchanging properties as long as the energy exchanges with its environment are much less than the difference ΔE between the first excited state and the ground state. It then acts like an "elementary" particle with fixed properties. Among those properties we mention its spatial extension and symmetry, its angular momentum, and its magnetic and/or electric multipole moments. The system changes these properties only if it is excited to higher states; whenever it returns to the ground state the system regains the properties typical of that state. Thus quantum mechanics introduces a "morphic" trait into the description of nature, with characteristic shapes, patterns, and symmetries. The atomic world abounds in such characteristic forms, typical properties, and ever-recurring qualities—from the identity of all atoms of a given kind to the faithful reproduction of living species.

Clearly, the Boltzmann paradox is resolved by the existence of discrete energy states. The states whose energies lie above the ground state by much more than kT cannot be excited when the system is in thermal equilibrium at the temperature T. Thus the excitation of the internal dynamics at a given temperature is limited. The internal degrees of freedom of the atom which give rise to quantum states whose energy is much higher than kT above the ground state do not reveal their existence in thermal equilibrium.

Another consequence of the discrete energy spectrum is that the quantum mechanical description of a system depends, in general, on the energy or temperature regime of interest. The wave function of a gas of N helium atoms at room temperature depends on $3N$ spatial coordinates. Each atom behaves like an elementary particle fully specified by its coordinates. On the surface of a star, that would be an inadequate description, because atomic excitation and ionization are commonplace. The wave function must now specify the position of the helium nuclei, and of the electrons. Deep in that star's interior, nuclear reactions occur, and there the appropriate wave function must keep track of the protons and neutrons that constitute the helium nuclei. In short, the Hilbert space of a system $\mathcal{S}$ depends on the phenomena of interest unless, of course, $\mathcal{S}$ is "truly elementary." One of our major goals is to determine which objects, if any, still survive this criterion of elementarity.

The quantization of energy for bound systems disposes of many facets of the Boltzmann paradox. But as we saw in §A, there are also aspects of the paradox that concern the electromagnetic field itself. These, too, are resolved by quantum mechanics. The quantum theory of the electromagnetic field is certainly not part of nonrelativistic quantum mechanics, and we shall deal with it at length. Here we confine ourselves to stating some essential facts that will be established in §C and Vol. II, Chap. II.

The stationary states of the electromagnetic field can be fully specified in terms of *photons*, or light quanta. Aside from the vacuum, which contains no photons, the simplest stationary states are those that contain one photon.

Such a state has a momentum $\mathbf{p}$ and energy E related by $p = E$(or $pc = E$ when $c \neq 1$). The photon therefore moves like a particle of zero mass. The wave function of a one-photon state has a space-time behavior that is characterized by a wave number $\mathbf{k}$ and frequency ω that are related to $\mathbf{p}$ and E by the deBroglie relation:

$$\mathbf{p} = \hbar\mathbf{k}, \qquad E = \hbar\omega. \tag{12}$$

An arbitrary stationary state of the field is then specified by stating the number $n_{\mathbf{k}}$ of photons of momentum $\mathbf{k}$ that are present (we now revert to natural units). The energy eigenvalue E_{tot} of the field is then

$$E_{\text{tot}} = \sum_{\mathbf{k}} n_{\mathbf{k}}|\mathbf{k}|. \tag{13}$$

When the quantized radiation field is at thermal equilibrium at some temperature T, the numbers $n_{\mathbf{k}}(T)$ are described by the Planck distribution.

These properties of the quantized electromagnetic field remove the remaining classical paradoxes. When the temperature of an object is raised, ever higher states are excited, and these emit photons of higher energy, or, put classically, the object emits light of shorter wavelength. Consequently the color shifts from red to white, the latter being due to the tail of the Planck distribution when its peak lies above the visible.

3. Rotations and angular momentum

The physical properties of any isolated system $\mathcal{S}$ are independent of its orientation. This symmetry has many important consequences for the system's level structure, as well as in collision and decay processes. Formally this symmetry is expressed by using three operators J_x, J_y, J_z, which are generators of infinitesimal rotations about the respective coordinate axes. Operators having precisely the same mathematical properties as the J's occur throughout particle physics, even though they have nothing to do with ordinary space. An understanding of rotations at a deeper level than is customary in elementary quantum mechanics is therefore indispensable.

The J_i also have a profound dynamical significance: they are the operators that represent the angular momentum observables. (To be precise, $\hbar\mathbf{J}$ is the angular momentum.) Indeed, it is generally true that the generators of infinitesimal geometrical transformations have such a dynamical role. The most familiar example is the linear momentum $\mathbf{p}$, which is the generator of infinitesimal spatial translations. Its relation to translations is particularly transparent when $\mathcal{S}$ is a single particle, for then $p_i = -i\partial/\partial x_i$, and the change of a function $g(x_i)$ under a translation through ε is just $\delta g = i\varepsilon p_i g$.

The linear momentum generates an infinitesimal translation, no matter what $\mathcal{S}$ may be—whether it is a single electron, or a complex structure held together by force fields. Angular momentum has the same universal relation to rotations, but it comes in many dynamical guises. In the case of the He atom, for example, each electron has its orbital angular momentum about the nucleus. But an electron *at rest* also has an angular momentum, called its *spin* or *intrinsic angular momentum*.

The term "spin" tends to have a usage that is as slippery as the term "elementary particle," and for somewhat similar reasons. Usually it is meant to be the angular momentum of an "elementary" particle as seen in its rest frame. But occasionally we say that "the spin of the nucleus is $\frac{3}{2}$," even though we know that the nucleus is made up of neutrons and protons, which have both orbital and spin angular momenta. When this language is used, one is usually dealing with a phenomenon where the object whose "spin" is under discussion behaves as if it were an "elementary" particle.

The bewildering array of angular momenta to which we have alluded does not, however, enter into the analysis of the properties of the operators J_i and their eigenstates. Because of their geometrical significance, these properties are universal—they do not depend on the nature of the system in question.

(a) The rotation operators

Let $\{|\psi_i\rangle\}$ be any set of states pertaining to $\mathcal{S}$ as seen by an observer using the coordinate frame F. Consider a frame F' rotated with respect to F. For each $|\psi_i\rangle$ there exists a state $|\psi_i'\rangle$ which, to the observer F', has precisely the same properties as $|\psi_i\rangle$ has when observed by F. When viewed from F, $|\psi_i'\rangle$ will be seen as a rotated replica of $|\psi_i\rangle$. Therefore $|\psi_i'\rangle$ and $|\psi_i\rangle$ are related by a unitary operator U:

$$|\psi'\rangle = U|\psi\rangle. \tag{14}$$

U depends on the relative orientation of F and F'. If F' is obtained from F by infinitesimal rotations $\delta\alpha$, $\delta\beta$, $\delta\gamma$ about the x-, y-, and z-axes, then

$$U = 1 - i(J_x\,\delta\alpha + J_y\,\delta\beta + J_z\,\delta\gamma). \tag{15}$$

If an observable A has an expectation value $\langle\psi|A|\psi\rangle$ that does not depend on the orientation of the arbitrary state $|\psi\rangle$, one calls A a scalar. Since, by hypothesis, $\langle\psi'|A|\psi'\rangle = \langle\psi|A|\psi\rangle$, where $|\psi'\rangle$ is given by (14), it follows that for a scalar observable $AU = UA$. In particular, the Hamiltonian H of our isolated system must be a scalar, and from $UH = HU$ it follows that

$$[J_i, H] = 0, \tag{16}$$

where $i = x$, y, or z. Furthermore, $UH = HU$ also implies that if $|\psi_E\rangle$ is an eigenstate of H, any rotated state $U|\psi_E\rangle$ is also an eigenstate with the same eigenvalue E.

Equation (15) only holds to first order in the infinitesimal angles. If one goes beyond that and studies successive rotations $F \to F' \to F''$, generated by the unitary operator $U_2 U_1$, one finds by a straightforward geometrical argument that $U_1 U_2 \neq U_2 U_1$, unless both rotations are about a common axis. These geometrical considerations show that the generators must satisfy the commutation rule

$$[J_x, J_y] = iJ_z \quad \text{(and cyclic permutations).} \tag{17}$$

As we have said, the structure of $\mathscr{S}$ never enters into the derivation of (17); these commutation rules follow inexorably from the properties of three-dimensional Euclidean geometry.

(b) Angular momentum eigenstates

The actual form of the angular momentum operators depends on the nature of $\mathscr{S}$, and on the choice of coordinates, but one rarely needs these explicit forms.

Since the J_i do not commute, a state can only be an eigenstate of one of the three operators. On the other hand, the operator

$$J^2 = J_x^2 + J_y^2 + J_z^2$$

is a scalar; hence it commutes with all three Js. One therefore chooses as eigenstates of the angular momentum simultaneous eigenstates of J^2 and, say, J_z. These states are characterized by two quantum numbers j and m. They are defined by

$$J^2|jm\rangle = j(j + 1)|jm\rangle,$$
$$J_z|jm\rangle = m|jm\rangle; \tag{18}$$

j and m are either integer or half-integer, and the quantum number m runs from $-j$ to $+j$ in unit steps:

$$m = -j, -j + 1, \ldots, +j - 1, +j.$$

Thus $(j(j + 1))^{\frac{1}{2}}$ is the eigenvalue of the angular momentum,[*] and m is the eigenvalue of its z-component. The operators J_x and J_y have the following effect on the states $|jm\rangle$:

$$(J_x \pm iJ_y)|jm\rangle = \sqrt{j(j + 1) - m(m \pm 1)} \, |jm{\pm}1\rangle. \tag{19}$$

As expected, the coefficients on the right-hand side of (19) do not depend on

---

[*] One often uses the abbreviation "angular momentum j," or "spin j." What is meant is the quantum number j.

the nature of the system. We shall use the following symbols for these operators which raise or lower m without changing j:

$$J_x + iJ_y \equiv J_+,$$
$$J_x - iJ_y \equiv J_- = J_+^\dagger. \tag{20}$$

All eigenstates of the Hamiltonian H of an isolated system must be expressible as eigenstates of J^2 and J_z, since these two operators commute with H. The two operators (20) are combinations of rotations, and as we have seen, their effect on the eigenstates of the Hamiltonian must lead to states of the same energy. Hence the $(2j + 1)$ states with the same j are degenerate and form a multiplet of states of the same energy and the same angular momentum j. Each substate of a multiplet represents the same state but with a different orientation in space. The degeneracy of these $(2j + 1)$ states is an expression of their indifference to that orientation. The degeneracy is lifted if, say, a magnetic field is introduced, for then the spatial symmetry is destroyed.

(c) Angular momentum $j = 1$: vectors

We shall treat two angular momentum multiplets explicitly, the first being the $j = 1$ triplet $|1m\rangle$, ($m = 0, \pm 1$). Its behavior under rotations is identical to that of the ordinary *real* 3-vector, $(V_x, V_y, V_z) = \mathbf{V}$. When rotated through $\delta\gamma$ about the z-axis, the change of $\mathbf{V}$ is

$$\delta V_x = -V_y \delta\gamma \qquad \delta V_y = V_x \delta\gamma \qquad \delta V_z = 0. \tag{21}$$

This rotation mixes x- and y-components, but one can choose another basis for $\mathbf{V}$ so that a z-rotation does not produce such mixing. For this purpose define*

$$V_{\pm 1} = \mp \frac{1}{\sqrt{2}} (V_x \pm iV_y). \tag{22}$$

From (21) we see that

$$\delta V_{\pm 1} = \pm iV_{\pm 1} \, \delta\gamma. \tag{23}$$

Now we compare this to the transformation of the $j = 1$ triplet, which we abbreviate by $|m\rangle$, with $m = \pm 1, 0$. In the Hilbert space** $\mathscr{C}_3$ of these states

---

* The factor (∓ 1) is a consequence of the phase convention adopted in (19).

** Arbitrary linear combinations of the $j = 1$ triplet, with *complex* coefficients, constitute the three-dimensional complex vector space $\mathscr{C}_3$, whereas the real vectors $\mathbf{V}$ span the Euclidean 3-space $\mathscr{E}_3$, which is a subspace of $\mathscr{C}_3$. Rotations are a transformation in $\mathscr{E}_3$. Consequently (15) does not generate the most general infinitesimal unitary transformation in $\mathscr{C}_3$. As we shall see in §3(e), in the special case of $j = \frac{1}{2}$ there is a correspondence between arbitrary unitary transformations in $\mathscr{C}_2$ and spatial rotations.

the angular momentum operators are the matrices defined by (18) and (19):

$$J_+ = \begin{pmatrix} 0 & \sqrt{2} & 0 \\ 0 & 0 & \sqrt{2} \\ 0 & 0 & 0 \end{pmatrix} \quad J_- = \begin{pmatrix} 0 & 0 & 0 \\ \sqrt{2} & 0 & 0 \\ 0 & \sqrt{2} & 0 \end{pmatrix} \quad J_z = \begin{pmatrix} 1 & 0 & 0 \\ 0 & 0 & 0 \\ 0 & 0 & -1 \end{pmatrix}.$$

Using this in (15), with $\delta\alpha = \delta\beta = 0$, the change of $|m\rangle$ is

$$\delta|m\rangle = -iJ_z\,\delta\gamma|m\rangle = -im|m\rangle\,\delta\gamma. \tag{24}$$

On comparing with (21) and (23), we see that the $(V_{+1}, V_z,$ and $V_{-1})$ components of $\mathbf{V}$ transform precisely like the $m = 1, 0,$ and -1 members of the $j = 1$ triplet, respectively.

(d) Angular momentum $j = \frac{1}{2}$: spinors

The simplest nontrivial angular momentum multiplet is the $|\frac{1}{2}m\rangle$ doublet, with $m = \pm\frac{1}{2}$. In the two-dimensional Hilbert space $\mathscr{C}_2$ spanned by these states, the operators J_i are 2×2 matrices. They are most conveniently expressed in terms of the *Pauli matrices*

$$J_i = \tfrac{1}{2}\sigma_i. \tag{25}$$

From (18) and (19) one finds

$$\sigma_x = \begin{pmatrix} 0 & 1 \\ 1 & 0 \end{pmatrix} \quad \sigma_y = \begin{pmatrix} 0 & -i \\ i & 0 \end{pmatrix} \quad \sigma_z = \begin{pmatrix} 1 & 0 \\ 0 & -1 \end{pmatrix}; \tag{26}$$

also

$$\sigma_+ = \tfrac{1}{2}(\sigma_x + i\sigma_y) = \begin{pmatrix} 0 & 1 \\ 0 & 0 \end{pmatrix},$$
$$\sigma_- = \tfrac{1}{2}(\sigma_x - i\sigma_y) = \begin{pmatrix} 0 & 0 \\ 1 & 0 \end{pmatrix}, \tag{27}$$

and therefore $J_\pm = \sigma_\pm$. Their commutation rules are

$$\sigma_i\sigma_j = -\sigma_j\sigma_i \qquad (i \neq j),$$
$$= 1 \qquad\qquad (i = j), \tag{28}$$
$$\sigma_x\sigma_y = i\sigma_z \text{ (and cyclic perms.).} \tag{29}$$

The properties (28) are specific to $j = \frac{1}{2}$, and do not apply to $j > \frac{1}{2}$.

Any state in $\mathscr{C}_2$ is a linear combination of the $m = \pm\frac{1}{2}$ eigenstates, and is called a *spinor*. Spinors, in one guise or another, crop up in all parts of our subject: in atomic, nuclear, and subnuclear physics. For that reason we shall discuss their properties in considerable detail.

As basis vectors in $\mathscr{C}_2$ we choose the $m = \frac{1}{2}$ and $m = -\frac{1}{2}$ states, and call them $|u\rangle$ and $|d\rangle$, respectively, where u stands for spin "up" along the z-direction, and d for "down." They are eigenstates of σ_z:

$$\sigma_z |u\rangle = |u\rangle, \quad \sigma_z |d\rangle = -|d\rangle. \tag{30}$$

Since $\langle u|u\rangle = \langle d|d\rangle = 1$, and $\langle u|d\rangle = 0$, we can also write them as 2-vectors:

$$|u\rangle = \begin{pmatrix} 1 \\ 0 \end{pmatrix} \quad |d\rangle = \begin{pmatrix} 0 \\ 1 \end{pmatrix}. \tag{31}$$

An arbitrary spinor $|\Phi\rangle$ is then

$$|\Phi\rangle = c_u |u\rangle + c_d |d\rangle, \tag{32}$$

where the complex numbers $c_{u,d}$ satisfy

$$|c_u|^2 + |c_d|^2 = 1. \tag{33}$$

The raising and lowering operators $\sigma_\pm$ have simple effects on the basis states:

$$\sigma_+ |d\rangle = |u\rangle \quad\quad \sigma_+ |u\rangle = 0, \tag{34}$$
$$\sigma_- |u\rangle = |d\rangle \quad\quad \sigma_- |d\rangle = 0, \tag{35}$$

as one sees from (27).

The behavior of spinors under rotation is of great importance. From (15) we know that a rotation can be represented by a 2×2 unitary matrix. Consider first an infinitesimal rotation $\delta\theta$ about the z-axis; from (15) and (25) we see that this is given by the matrix

$$U = 1 - \tfrac{1}{2} i \sigma_z \, \delta\theta. \tag{36}$$

A finite rotation through θ can be viewed as n successive rotations through the angle θ/n, and if $n \to \infty$, these are all infinitesimal. Hence this finite rotation is given by the matrix

$$U(\theta) = \lim_{n\to\infty} \, [1 - \tfrac{1}{2} i \sigma_z (\theta/n)]^n \tag{37}$$
$$= \exp(- \tfrac{1}{2} i \sigma_z \theta). \tag{38}$$

Since $\sigma_z^2 = 1$, [see (28)], this can also be written as

$$U(\theta) = \cos \tfrac{1}{2}\theta - i\sigma_z \sin \tfrac{1}{2}\theta. \tag{39}$$

Because space is isotropic, the choice of z-axis is arbitrary. We could just as well have rotated through θ about an axis having an arbitrary orientation specified by the unit vector $\hat{\mathbf{n}}$. The generator of rotations is then $\hat{\mathbf{n}} \cdot \mathbf{J}$ quite generally, and $\mathbf{J} = \frac{1}{2}\boldsymbol{\sigma}$ when $j = \frac{1}{2}$, where $\boldsymbol{\sigma} = (\sigma_x, \sigma_y, \sigma_z)$. Consequently the transformation of a spinor under the most general rotation, as specified by an axis and angle of rotation $(\hat{\mathbf{n}}, \theta)$, is given by $|\Phi'\rangle = U|\Phi\rangle$, where

$$U(\hat{\mathbf{n}}, \theta) = \cos \tfrac{1}{2}\theta - i\hat{\mathbf{n}} \cdot \boldsymbol{\sigma} \sin \tfrac{1}{2}\theta. \tag{40}$$

This matrix depends on three angles: θ and the two angles that specify $\hat{\mathbf{n}}$. The arbitrary spinor $|\Phi\rangle$, defined in (32), also depends on three real parameters, because the two complex numbers c_u and c_d are constrained by (33). It is therefore not surprising that $|\Phi\rangle$ can always be written as some rotation applied to either one of the basis spinors $|u\rangle$ or $|d\rangle$. Conversely, for any $|\Phi\rangle$ there always exists a spatial coordinate frame F such that $|\Phi\rangle$ is either an "up" or a "down" spin state with respect to F's z-axis. (To prove these statements we merely apply U to, say $|u\rangle$, read off the resulting coefficients of $|u\rangle$ and $|d\rangle$, and equate them to c_u and c_d.) We may therefore call its spin direction the direction along which the arbitrary state $|\Phi\rangle$ is an "up" state.

The rotation matrix (40) has a remarkable property: If one rotates through $\theta = 2\pi$ about any axis, i.e., returns to the original spatial coordinate system, one finds $U = -1$, not $U = 1$, as one would have expected! In other words, the state vector of a $j = \frac{1}{2}$ system is double-valued in ordinary 3-space, and returns to itself only after a rotation through 4π. That is why spinors do not arise in classical physics, where all quantities must be single-valued. In quantum mechanics, however, the overall phase of a probability amplitude is not observable, and spinors are therefore admissible. On the other hand, relative phases are meaningful, and this double-valuedness has been established[*] by studying the interference of two beams composed of $j = \frac{1}{2}$ particles (neutrons). One observes a maximum phase shift in two coherent neutron beams when the spin of only one beam is turned through 360° by means of a precession of the magnetic moment in a suitable magnetic field. A second turn by 360° reestablishes the original phase!

(e) Two-level systems and the spin analogy

If we reexamine our analysis of the $j = \frac{1}{2}$ doublet, we will note that all that really mattered was that we were dealing with the two-dimensional Hilbert space $\mathscr{C}_2$. That $|u\rangle$ and $|d\rangle$ were angular momentum states was irrelevant, because all our results depend only on the fact that (32) is the most general state in $\mathscr{C}_2$, provided the restriction (33) is satisfied. This leads us to the following remark:

Whenever a system (or subsystem) $\mathscr{S}$ can be described by just two orthogonal state vectors $|\alpha\rangle$ and $|\beta\rangle$, all observables pertaining to $\mathscr{S}$ can be put into

---

[*] Werner (1980).

one-to-one correspondence with some linear combination of the Pauli matrices and the 2×2 unit matrix. This remark is exploited in several different contexts in particle physics, e.g., in isospin. The word "spin" crops up in these contexts because of the underscored statement. Put equivalently, to every unitary transformation in the space $\mathscr{C}_2$ spanned by $|\alpha\rangle$ and $|\beta\rangle$ there corresponds a rotation in an Euclidean 3-space $\mathscr{E}_3$. If $|\alpha\rangle$ and $|\beta\rangle$ have nothing to do with angular momentum, the rotations in $\mathscr{E}_3$ have nothing to do with ordinary spatial rotations. It should be stressed that this analogy to spin $\frac{1}{2}$ does not require that $|\alpha\rangle$ and $|\beta\rangle$ have the same energy; for that matter, even in the case of a "true" spin the levels can be split by a magnetic field. On the other hand, if the Hamiltonian H of $\mathscr{S}$ is invariant under "rotations" in $\mathscr{C}_2$, $|\alpha\rangle$ and $|\beta\rangle$ are degenerate. H has this property if it commutes with the Pauli matrices that act on $\mathscr{C}_2$.

(f) The addition of angular momenta

Let $\mathbf{J}^{(1)}$ and $\mathbf{J}^{(2)}$ be two dynamically independent, or commuting, angular momenta. They could refer to two distinct systems, such as the two electrons in the He atom, or to a single electron's orbital and intrinsic angular momenta. Frequently one needs the eigenstates of the total angular momentum

$$\mathbf{J} = \mathbf{J}^{(1)} + \mathbf{J}^{(2)}. \tag{41}$$

This is called the problem of angular momentum addition.

The component angular momenta $\mathbf{J}^{(1)}$ and $\mathbf{J}^{(2)}$ provide us with a complete set of eigenstates $\{|j_1 m_1 j_2 m_2\rangle\}$, which are products of the states $|j_1 m_1\rangle$ and $|j_2 m_2\rangle$. The sought-after states are some linear combination of this complete set. Note that the product states are simultaneous eigenstates of four operators: $(\mathbf{J}^{(i)})^2$ and J_z^i, for $i = 1$ and 2. The new states will be eigenstates of the four operators $\mathbf{J}^2, J_z, (\mathbf{J}^{(1)})^2, (\mathbf{J}^{(2)})^2$, which commute among themselves, because $(\mathbf{J}^{(i)})^2$ is a scalar, and like any scalar commutes with the total angular momentum. Consequently the new eigenstates also possess the quantum numbers j_1 and j_2; they will be written as $|j_1 j_2 jm\rangle$, where j and m are the eigenvalues associated with $\mathbf{J}$. Since $J_z = J_z^{(1)} + J_z^{(2)}$,

$$m = m_1 + m_2. \tag{42}$$

All that remains unknown are the so-called Clebsch–Gordan coefficients $\langle j_1 m_1 j_2 m_2 | jm\rangle$ in the linear transformation

$$|j_1 j_2 jm\rangle = \sum_{m_1 m_2} |j_1 m_1 j_2 m_2\rangle\langle j_1 m_1 j_2 m_2 | jm\rangle \tag{43}$$

where the sum is restricted by (42). Given j_1 and j_2, the total angular momentum is constrained by the triangular inequality $|j_1 - j_2| \leqslant j \leqslant j_1 + j_2$, or

$$j = |j_1 - j_2|, |j_1 - j_2| + 1, \ldots, j_1 + j_2, \tag{44}$$

and of course $m = -j, -j + 1, \ldots, j$.

The simplest example of angular momentum addition is provided by $j_1 = j_2 = \frac{1}{2}$. We shall discuss this case explicitly, as it arises frequently. There are altogether four states, $|u_1u_2\rangle$, $|d_1d_2\rangle$, $|u_1d_2\rangle$, and $|u_2d_1\rangle$, where $|u_i\rangle$ and $|d_i\rangle$ are the spinors (31), and the subscript 1 or 2 tells us to which angular momentum they refer. Obviously the "up-up" state has the eigenvalue $m = 1$ of J_z, "down-down" has $m = -1$, whereas the others are $m = 0$ states. For any angular momentum j, no matter what its origin, *all* m values $m = j, m = j - 1, \ldots, m = -j$, must be present in the multiplet. Consequently $|u_1u_2\rangle$ and $|d_1d_2\rangle$ are $j = 1$ states. Some linear combination of the remaining states $|u_1d_2\rangle$ and $|u_2d_1\rangle$ must provide the $m = 0$ partner of $|u_1u_2\rangle$ and $|d_1d_2\rangle$, leaving the other orthogonal combination that can only be $j = 0$, $m = 0$. This conforms with the general result (44). To determine these two $m = 0$ states, we note that (19) tells us how to build a $j = 1, m = 0$ state from $j = 1, m = 1$: we need merely apply the operator J_- to the latter. But $J_- = J^{(1)}_- + J^{(2)}_-$ is symmetric in the labels 1 and 2. Since $|u_1u_2\rangle$ is already symmetric, $J_-|u_1u_2\rangle \propto |u_1d_2\rangle + |u_2d_1\rangle$. The normalized $j = 1$ multiplet, often called the "parallel spin" states, are therefore

$$|11\rangle = |u_1u_2\rangle,$$

$$|10\rangle = \frac{1}{\sqrt{2}}[|u_1d_2\rangle + |u_2d_1\rangle], \tag{45}$$

$$|1 - 1\rangle = |d_1d_2\rangle.$$

The $j = m = 0$ "antiparallel" state must be orthogonal to $|10\rangle$, and is therefore

$$|00\rangle = \frac{1}{\sqrt{2}}[|u_1d_2\rangle - |u_2d_1\rangle]. \tag{46}$$

If the dynamical system described by the states (45) and (46) has a rotationally invariant Hamiltonian H, it will have energy eigenstates that can be labelled by j and m. The $j = 1$ triplet will be degenerate, but in general the $j = 0$ singlet will have a different energy. Insofar as angular momentum is concerned, H has a very simple form. It can be some function of $\boldsymbol{\sigma}^{(1)}$ and $\boldsymbol{\sigma}^{(2)}$, the Pauli matrices associated with the two independent angular momenta. Since H is rotationally invariant, it is only a function of $(\boldsymbol{\sigma}^{(1)})^2$, $(\boldsymbol{\sigma}^{(2)})^2$, and $\boldsymbol{\sigma}^{(1)} \cdot \boldsymbol{\sigma}^{(2)}$. But

$$(\boldsymbol{\sigma})^2 = 3 \tag{47}$$

for any Pauli spin, as we see from (28), and is a pure number. Furthermore, (29) says that any product of Pauli matrices can be reduced to a linear combination of such matrices. Hence the most general operator is only linear in Pauli matrices; (39) and (40) illustrate this. Consequently H has the general form

$$H = A + B\,\boldsymbol{\sigma}^{(1)} \cdot \boldsymbol{\sigma}^{(2)}, \tag{48}$$

where A and B are operators (or just numbers) that do not involve the σ's, but could, for example, depend on spatial coordinates.

To evaluate the eigenvalues of the operator in (48), note that

$$(\mathbf{J})^2 = \tfrac{1}{4}[\boldsymbol{\sigma}^{(1)} + \boldsymbol{\sigma}^{(2)}]^2 = \tfrac{1}{2}\boldsymbol{\sigma}^{(1)} \cdot \boldsymbol{\sigma}^{(2)} + \tfrac{3}{2},$$

where (47) was used; but the left-hand side is $j(j + 1) = 0$ or 2, and therefore

$$\boldsymbol{\sigma}^{(1)} \cdot \boldsymbol{\sigma}^{(2)} = \begin{cases} 1 & (j = 1) \\ -3 & (j = 0) \end{cases} \tag{49}$$

The spectrum of H is therefore split by a term that depends on the total angular momentum, or equivalently, by a term that depends on the symmetry of the state under interchange of the constituents 1 and 2.

What about combining three spinors? Here we will not go into as much detail as in the case of two. Suffice it to say that we can treat this case by first combining two, followed by one additional spinor. The first two give us states with angular momenta $j = 1$ and 0. Combining the $j = 1$ state with an angular momentum $j = \tfrac{1}{2}$ gives us $J = \tfrac{3}{2}$ and $\tfrac{1}{2}$; combining $j = 0$ with $j = \tfrac{1}{2}$ gives $J = \tfrac{1}{2}$. Altogether we therefore get one $J = \tfrac{3}{2}$ state and two $J = \tfrac{1}{2}$ states:

$$\underline{(j = \tfrac{1}{2})^2} \qquad \underline{(j = \tfrac{1}{2})^3}$$

$$\uparrow \qquad\qquad \uparrow$$
$$\uparrow \quad \uparrow\downarrow \qquad \uparrow \quad \uparrow$$
$$\qquad\qquad \uparrow \quad \uparrow\downarrow \quad \uparrow\downarrow\uparrow$$
$$J: \quad 1 \quad 0 \qquad J: \quad \tfrac{3}{2} \quad \tfrac{1}{2} \quad \tfrac{1}{2}$$

4. Space reflection and parity

Rotations are examples of continuous transformations, because any finite rotation can be viewed as an infinite sequence of infinitesimal rotations; recall Eq. (37). There are also *discrete* transformations; these cannot be built up by a succession of small steps. The most familiar discrete transformation is a *reflection in a plane*, say the x–y plane:

$$(x,y,z) \rightarrow (x,y,-z). \tag{50}$$

A *reflection through the origin*,

$$(x,y,z) \rightarrow (-x,-y,-z) \tag{50'}$$

is obtained from (50) by a rotation through 180° about the z-axis. It is therefore a matter of taste whether one uses reflection through a plane or the origin when one has rotational symmetry: both are transformations that are *not* equivalent to *any* rotation.

We now address ourselves to the question of whether some set of physical phenomena imply that the underlying laws are reflection invariant. Note the distinction we have drawn between phenomena and laws. Most laws of physics have a high degree of symmetry, but allow motions that do not exhibit that symmetry. For example, the wave equation of acoustics is invariant under rotations, and yet a spherically symmetric sound wave is a rarity. For that reason it is often difficult to extract the underlying symmetry of a physical law from the phenomena it describes. A symmetric law implies sets of phenomena that are related to each other by the symmetry operation. In the acoustical example, the rotational invariance of the equation implies that if, say, there is a certain wave packet traveling in some direction, other packets exist that are identical in every respect except that they propagate in any other direction.

Let us return to the problem of interest, reflections. Quite generally, *a law of nature is said to be reflection invariant if the probability for any process equals the probability for the mirror image of that process.* Application of this criterion will be found in §§D.4 and E.9(f). Here we illustrate it with an example from atomic physics. Take an ensemble of atoms, all in a definite excited state, and consider the angular distribution of the emitted radiation. This distribution will be anisotropic if the excited state is anisotropic, e.g., a definite substate m of a $j \neq 0$ multiplet. If the laws that govern the structure of the atoms, *and* the emission of photons, are *reflection or mirror invariant*, the angular distribution must be such that it is unchanged if it is reflected in a mirror which leaves the excited state unchanged. For the state $|jm\rangle$ this is the case if the surface of the mirror is perpendicular to the z-axis, because a rotation in the x–y plane is unaffected by a reflection through that plane.

The only laws of nature known to us which do not display reflection symmetry govern the so-called weak interactions. As the name implies, their effects are weak in our presently known realm of phenomena. If these "weak" effects are neglected, the mirror symmetry of all the other interactions gives rise to a quantum number called *parity*.

We introduce a *space reflection operator P* which, when acting on any state $|\psi\rangle$, produces another state $|\psi'\rangle$ wherein all spatial coordinates are inverted: $P|\psi\rangle = |\psi'\rangle$. Since two reflections bring us back to the original state, $P^2 = 1$. Consequently P has only two eigenvalues, $\Pi = \pm 1$. These eigenvalues are called the parity of the state in question. Eigenstates of P are called even or odd according to whether $\Pi = 1$ or -1.

We will now show that when mirror symmetry holds, it is possible to ascribe a parity to stationary states. To put it more accurately, mirror symmetry implies that nondegenerate stationary states are necessarily eigenstates of P, while in the case of degenerate states, one can always choose an orthogonal set among the degenerate states such that they are eigenstates of P. Let $|\psi_E\rangle$ be an eigenstate of H. Mirror symmetry requires $P|\psi_E\rangle$ to be a state with the same eigenvalue E of H. If there is *no* degeneracy, $P|\psi_E\rangle$ can only be the sole state $|\psi_E\rangle$ itself, apart from an overall factor:

$$P|\psi_E\rangle = a|\psi_E\rangle.$$

Thus $|\psi_E\rangle$ is an eigenstate of P, and from our preceding argument, the eigenvalue $a = \Pi = \pm 1$ is its parity. If the level E is degenerate, $P|\psi_E\rangle$ can be a different state vector from $|\psi_E\rangle$, but we can then form the two combinations

$$|\psi_E^{\pm}\rangle = \frac{1}{\sqrt{2}}(|\psi_E\rangle \pm P|\psi_E\rangle), \tag{51}$$

each of which is an eigenstate of the same energy E. Clearly $|\psi_E^{+}\rangle$ has parity $+1$, and $|\psi_E^{-}\rangle$ has parity -1. Hence we conclude that when mirror symmetry holds all stationary states can be chosen to have a definite parity.

As an example, consider the eigenstates of two interacting particles without spin in field-free space, with their center-of-mass at rest. The wave function then depends only on their separation $\mathbf{x}$. The dependence on the orientation of $\mathbf{x}$ is expressed by a spherical harmonic $Y_{lm}(\theta,\phi)$, where l is the orbital angular momentum. The parity operation changes the polar angle θ into $\pi - \theta$, and the azimuth ϕ into $\phi + \pi$. But $Y_{lm}(\pi - \theta, \phi + \pi) = (-1)^l Y_{lm}(\theta,\phi)$, and therefore the parity is $(-1)^l$.

The parity is a *multiplicative quantum number*. If a system $\mathscr{S}$ is composed of two systems $\mathscr{S}_1$ and $\mathscr{S}_2$, having the parities Π_1 and Π_2, the parity Π of the combined system is $\Pi = \Pi_1 \cdot \Pi_2$. This is seen as follows. If $\mathscr{S}_1$ and $\mathscr{S}_2$ are far apart and noninteracting, the state vector describing the combined system is the product of the states of $\mathscr{S}_1$ and $\mathscr{S}_2$. Then the rule is obvious. If mirror symmetry holds, the parity remains the same in any dynamical process that brings the two subsystems together.

Most quantum numbers are not multiplicative. For example, the electric charge, or the "baryon number," which will be introduced later, are additive quantum numbers. Evidently the total charge and the total number of baryons when two bodies are combined is the sum of these numbers in each separate entity.

So far, we have talked of parity as a quantum number associated with the spatial wave function that describes a system of particles. We have not assigned any parity that is intrinsic to these constituents. Is there such a thing as an *intrinsic parity*? That there is can be seen from the following example.

Consider two different spin zero states of the He atom. The first, $|\psi_s\rangle$, has both electrons in the lowest s orbit, giving no orbital angular momentum L, and their spins antiparallel to total spin $S = 0$, and therefore $J = 0$. The other, $|\psi_p\rangle$, has one electron in a p orbit ($l = 1$), the other in the lowest s orbit, giving $L = 1$, and the two spins parallel, so that $S = 1$; L and S are then coupled to give $J = 0$. The parities of these states are $\Pi_s = +1$, and $\Pi_p = -1$.

These two $J = 0$ states $|\psi_s\rangle$ and $|\psi_p\rangle$ are examples of what are called *scalars* and *pseudoscalars*, respectively. A scalar is a quantity, like temperature or pressure, that does not depend on the orientation of the coordinate system, and is also reflection invariant, whereas a pseudoscalar shares this orientation invariance, but is odd under reflections.

A similar distinction applies to vectors. There are *polar vectors*, such as the position, the linear momentum, and the electric field $\mathbf{E}$, which change sign under reflection, and *axial vectors*, like angular momentum and the magnetic field $\mathbf{B}$, which do not change sign. Consequently $\mathbf{E} \cdot \mathbf{B}$ is an example of a pseudoscalar. One can also form the pseudoscalar $\mathbf{a} \cdot (\mathbf{b} \times \mathbf{c})$ from three polar vectors.

We return to our helium example, and consider two distinct states in which the center of mass motion of the He atom is described by the same wave function $\chi(\mathbf{R})$. In the first the He atom is in the internal state $|\psi_s\rangle$; in the second, it is in the internal state $|\psi_p\rangle$. If we think of $|\psi_s\rangle$ and $|\psi_p\rangle$ as elementary particles of spin zero, we see that even though their motion is described by identical wave functions χ, they have different properties when spatially reflected. Hence the "intrinsic" parities Π_s and Π_p are physically meaningful.

That being settled, one must ask how one is to determine the intrinsic parity of an object whose internal structure is unknown to us, and which may actually have no such structure. In atomic physics this question is easily answered, but in nuclear and particle physics it becomes a subtle issue to which we will return from time to time.

In any atomic process the total number of electrons, and of nuclei of various sorts, remains unchanged. Only the photon number changes. We consider only the most important example, where one photon is emitted or absorbed; the generalization to multiphoton processes is straightforward. The parity of the initial (or final) state is composed of three factors: (1) the product of all the intrinsic parities of the electrons and nuclei; (2) the orbital parity of the electronic wave function Π_i (or Π_f); (3) the parity Π_γ of the emitted or absorbed one-photon state. Reflection invariance requires the product of these three factors to be the same for the initial and final states, but the factor stemming from the intrinsic parities of electrons and nuclei is common, and drops out. In short, *the intrinsic parities of electrons and nuclei are arbitrary in atomic physics.* What parity conservation does require is

$$\Pi_i \Pi_f = \Pi_\gamma. \tag{52}$$

If the photon were a particle with non-zero mass, Π_γ would be the product of the photon's intrinsic parity, and the parity of its orbital wave function. But for a zero-mass particle, there is no clear-cut separation into orbital and intrinsic motion. Nevertheless, there is a simple solution to our problem, for, as we shall see in §II.A (Vol. II), there is a one-to-one correspondence between the wave functions of one-photon states, and the solutions of Maxwell's classical equations. In particular, the one-photon wave functions of definite total angular momentum and parity are identical to certain electric or magnetic multipole fields. Of these the most important are the electric and magnetic dipole fields, designated by $E1$ and $M1$, for which $\Pi_\gamma = -1$ and $+1$, respectively.

5. Fermions and bosons

One of the most important symmetries arises from the indistinguishability of particles of a given kind. For example, all electrons are completely identical in their properties and behavior. The dynamics of any system must be invariant under the interchange of one particle by another of the same kind. Quantum mechanics offers two possible realizations of this invariance. Since two successive interchanges of the same pair reestablish the original state, the only change in the state vector that one interchange can cause is multiplication by $+1$ or -1. Consequently the state of N identical (or indistinguishable) particles is either symmetric or antisymmetric under the interchange of any pair. Particles described by symmetric states are called *bosons*, those described by antisymmetric states *fermions*.*

Nonrelativistic quantum mechanics provides no clue as to whether a certain type of particle is a boson or fermion—this decision is left to experiment. We know experimentally that photons and π-mesons are bosons, and that electrons, protons, and neutrons are fermions. We also know that these bosons have integer spin (1 for the photon, zero for the pion), whereas all these fermions have spin $\frac{1}{2}$. Quantum field theory shows that this is no accident (see §C.6).

In many cases the state of a system of fermions is reasonably well described by the approximation that the constituent particles move independently of each other, and can therefore be considered to occupy single-particle states. Under this circumstance antisymmetry is equivalent to the requirement that not more than one fermion of the same kind occupies a given single-particle state, i.e., to the *Pauli Exclusion Principle*.

Another consequence of the antisymmetry of many-fermion states is that two identical fermions (for example, two electrons with the same spin orientation) cannot be found at the same spatial point. Since the relative

---

* These names refer to the statistical properties of an ensemble of many particles: ensembles of fermions are described by Fermi–Dirac statistics, of bosons by Bose–Einstein statistics.

momentum p is determined by the rate at which the wave function changes with separation, two identical fermions tend to stay away from each other at distances of the order of $1/p$ or larger. For bosons the opposite occurs.

It can easily be shown that a bound system of n fermions is itself a fermion if n is odd and a boson if n is even. This determines the statistics of atomic nuclei, which are bound systems of protons and neutrons. If the number of these constituents is odd, the nucleus behaves as a fermion; if even, as a boson. However, these statements are only true if the constituents remain in their ground state; should some constituents become excited, they would no longer be identical to those that are unexcited.

6. The electromagnetic nature of atomic phenomena

The most important consequence of the application of quantum mechanics to atomic systems is the recognition that all properties of atoms, molecules, and their aggregates, can be understood by assuming that an atom is a system consisting of a nucleus small compared to atomic dimensions with a charge Ze, and of Z electrons, each of charge $-e$, with the interaction between these constituents being solely due to the electromagnetic fields produced by the charges. This dynamical problem is simple in principle; it requires only non-relativistic quantum mechanics since the resulting velocities are, in general, much smaller than c. It is not simple in practice because the treatment of many particle systems presents many difficulties, some of which are not yet overcome. Nevertheless, we are certain that all the interatomic and inter-molecular forces listed in §A are manifestations of the electromagnetic interactions between the constituents, among which the electrostatic attraction or repulsion (Coulomb force) plays the dominant role. Since almost all natural phenomena that we observe in our terrestrial environment (including life) are due to interactions between atoms, we conclude that these phenomena are all consequences of the electromagnetic interaction between nuclei and electrons, and of quantum mechanics.

That electrons are fermions plays an absolutely crucial role in atomic physics, and *ipso facto*, in everyday life. The shell structure of atoms, and the associated periodic table of chemistry, are consequences of the Pauli principle. Bulk matter (and biological systems) could not exist in anything like the forms that we know were it not that electrons obey Fermi–Dirac statistics.

The energy spectra and bound state wave functions of atoms depend on only three parameters: the pure numbers Z and α, and the electron mass m. Z was just defined, and α is the *fine structure constant*

$$\alpha = \frac{e^2}{4\pi\hbar c} \simeq \frac{1}{137};$$

(53)

α is the crucial parameter that governs all electromagnetic phenomena when quantum effects are significant. All energies can be expressed in units of

$$m \simeq 0.511 \text{ MeV}, \tag{54}$$

and lengths in units of the electron Compton wavelength

$$\frac{1}{m} = 3.86 \times 10^{-11} \text{ cm}. \tag{55}$$

The nuclear mass is so large compared to m that it hardly plays a role in atomic structure.

The size of the hydrogen atom can be estimated by writing the energy in terms of the momentum and position uncertainties:

$$E \simeq \frac{1}{2m} (\Delta p)^2 - \frac{\alpha}{(\Delta x)}. \tag{56}$$

Since $\Delta p \sim 1/\Delta x$, we can consider this to be a function of Δx alone, and after minimizing we find that Δx is

$$a_0 = \frac{1}{\alpha m} = 0.529 \times 10^{-8} \text{ cm}. \tag{57}$$

This is the Bohr radius; it is the rough characteristic dimension of most atomic states.

According to (56), the characteristic binding energy is

$$\text{Ry} = \frac{\alpha}{2a_0} = \frac{1}{2} m \alpha^2 = 13.6 \text{ eV}, \tag{58}$$

and is called the Rydberg. All low-lying atomic excitations, which involve only electrons in the outermost shells, are of order Ry. As one sees from (58), the characteristic atomic velocity is α.

In the units that we shall use, the ground state of hydrogen has the energy

$$E_H = M + m - \frac{1}{2} m \alpha^2,$$

where $M = 1836m$ is the proton mass.

Atomic dimensions are very large compared to the electron's Compton wavelength, and atomic binding energies are far smaller than the electron's mass. These are the characteristics of a nonrelativistic system. *In general, a system that is large compared to the Compton wavelength of its constituents has binding energies small compared to their rest masses, and nonrelativistic internal velocities*. We shall exploit this important fact on several occasions. Its proof only requires the uncertainty principle and the virial theorem. The characteristic size R determines the characteristic velocity to be $v \sim 1/Rm$, so $v \ll 1$ if R is much larger than the Compton wavelength m^{-1}. According to the virial theorem, the mean total and kinetic energies are of the same order of magnitude, and therefore $E \sim mv^2 \ll m$ since $v \ll 1$.   Q.E.D.

An isolated atom or molecule emits a photon γ spontaneously when it is in a state above the ground state, and ends up in a lower state. The energy ω of the photon is equal to the energy difference ΔE between the states. The inverse transition occurs when a suitable photon is absorbed.* Quantum mechanics describes these processes as resulting from the coupling of the electric charges of the particles with the electromagnetic field.

The wavelength $\lambda = 2\pi/\omega$ of photons emitted by atoms—or by *any* nonrelativistic bound state—are long compared to the dimensions R of the emitter. This follows from our previous conclusion that for such a system $R \gg m^{-1}$. If we define $\lambdabar = \lambda/2\pi$, we have $\lambdabar/R = 1/R\Delta E$, and $\Delta E \sim 1/mR^2$, which gives $\lambdabar/R \sim Rm \sim 1/v \gg 1$. The radiation field of a nonrelativistic emitter can therefore be expanded in powers of the small parameters v and $R/\lambdabar$, which is also of order v. This is called the multipole expansion; the first two terms are the electric ($E1$) and magnetic dipole ($M1$) fields. Most visible atomic transitions are $E1$.

Angular momentum and parity conservation play an essential role in determining the character of a radiative transition. In emission, for example, the initial angular momentum J_i of the atom must be shared between that of the photon, J_γ, and of the final atomic state, J_f. Consequently (44) must be satisfied:

$$|J_f - J_\gamma| \leq J_i \leq J_f + J_\gamma. \tag{59}$$

In §C.2 we shall learn that a one-photon state with $J_\gamma = 0$ does *not* exist. Hence there are no electromagnetic one-photon transitions between spin zero states.

The one-photon states corresponding to dipole fields have $J_\gamma = 1$. In the case of dipole radiation (59) therefore reduces to the selection rule

$$\Delta J = 0 \text{ or } 1. \tag{60}$$

In §4 we derived the parity conservation law (52). The photon parities are -1 and $+1$ for $E1$ and $M1$ radiation, respectively. Consequently (60) must be supplemented by

$$\Pi_i = -\Pi_f \quad (E1),$$
$$\Pi_i = \Pi_f \quad\ (M1). \tag{61}$$

---

* Emission and absorption of two or more photons in one transition are possible, but the probability of this is very small.

The angular distribution of the emitted light depends only on the angular momentum quantum numbers of the initial and final states. When the direction of the initial and final angular momentum is randomly distributed, the distribution is necessarily isotropic.

The lifetime τ of excited atomic states is finite, since they emit light. This gives rise to a finite width $\Gamma = 1/\tau$ of each excited state. The coupling of the electron's motion with the radiation field allows one to calculate these widths. They are of order $f\alpha^3$ Ry, where the numerical coefficient f is of order one for the most common electric dipole transitions.

7. Free particle wave functions

Let $\mathscr{S}$ be any isolated object that is not subjected to forces. It may be considered "elementary," like the electron, or we may know that it is a complex atom or nucleus. $\mathscr{S}$ is specified by its nature (electron, proton, atom, nucleus) and by the state in which it happens to be (e.g., the third excited state). This information includes its energy when it is at rest, that is, its mass M. In order to completely describe the state of $\mathscr{S}$, four more quantum numbers are needed. Three of these quantum numbers are eigenvalues of observables that are familiar from classical mechanics: they can, for example, be the position $\mathbf{r}$ of the objects center of mass, or, as in Eq. (62), its total momentum $\mathbf{p}$.

The fourth quantum number describes the orientation of the system's intrinsic angular momentum $\mathbf{s}$—the spin. As long as we are concerned with nonrelativistic phenomena, we can use the definition of intrinsic angular momentum given on p. 16, i.e., the angular momentum observed in the rest frame F_0 of $\mathscr{S}$. The state of such a freely moving system is then represented by

$$\psi_m(\mathbf{p};\mathbf{r}t) = |sm\rangle \exp[i(\mathbf{p} \cdot \mathbf{r} - Et)], \tag{62}$$

where $|sm\rangle$ is an angular momentum state as defined by (18) with $s = j$, and m is the projection of $\mathbf{s}$ along the z-axis of F_0. The value of s, the spin of the system, is one of the fixed properties of $\mathscr{S}$, like its electric charge, and as it is not a variable, it is often supressed.

The simplest case is $s = 0$. In that case, (62) is often called a scalar wave since $|sm\rangle = |00\rangle$ transforms like a scalar. Under a spatial rotation such a state behaves just like an acoustical wave.*

The other case that is familiar to us is $s = 1$, because we have already seen that the three $s = 1$ states transform like a 3-vector. For that reason one can call (62) a vector wave when $s = 1$. It has three components. Elastic waves in a noncrystalline solid are an example. Electromagnetic waves (photons) are also an example, but as they are only transversely polarized, they are a very

---

* In carrying out a rotation of the one-particle state (62), one must remember that the momentum $\mathbf{p}$ is also to be rotated, not just the spin state.

special type of vector wave. Further discussion of the photon will be found in §C.2.

The third important case is spinor waves, when $s = \frac{1}{2}$. Nature provides us with many $s = \frac{1}{2}$ particles, and spinor waves are therefore of considerable importance.

Intrinsic angular momentum can also be defined without going to the rest frame of $\mathcal{S}$. We will need such a definition for relativistic situations, especially when $\mathcal{S}$ has no mass (e.g., a photon) for then there is no rest frame. Therefore we need a frame-independent definition of spin. This is provided by the *helicity operator* Λ, which is defined to be the projection of the total angular momentum $\mathbf{J}$ along the direction $\hat{\mathbf{p}}$ of the linear momentum:

$$\Lambda = \mathbf{J} \cdot \hat{\mathbf{p}}. \tag{63}$$

The eigenvalues h of Λ are called the *helicity*. The spin s of $\mathcal{S}$ turns out to be the largest eigenvalue h_{max}. This can be seen as follows. The orbital part $\mathbf{L}$ of $\mathbf{J}$ does not contribute to Λ, because $\mathbf{L}$ is always perpendicular to the momentum; hence $\Lambda = \mathbf{s} \cdot \hat{\mathbf{p}}$. To evaluate the eigenvalues h, we observe that Λ generates rotations about $\hat{\mathbf{p}}$. Lorentz transformations along an axis, and rotations about that axis, commute. Hence Λ is unchanged by a transformation along $\hat{\mathbf{p}}$ to a frame F in which the particle moves slowly.* But nonrelativistic quantum mechanics is applicable in F, whence $h = s$, $s - 1, \ldots, -s$.

Note that Λ is a pseudoscalar, i.e., invariant under rotations, but odd under reflection. Consequently h can be specified simultaneously with J^2, while $h \to -h$ under reflection.

Finally, we point out that the product form (62) only holds for non-relativistic motions. When v is comparable to c, it is still possible to write the wave function as a product of $\exp[i\mathbf{p} \cdot \mathbf{r}]$ with another factor that describes the spin, but the spin function then depends on the momentum as well. This is actually familiar from classical electrodynamics. A running plane wave is described by the vector potential

$$\mathbf{A} = A_0 \hat{\boldsymbol{\varepsilon}}_{\mathbf{k}} \exp[i(\mathbf{k} \cdot \mathbf{r} - \omega t)], \tag{64}$$

where $\hat{\boldsymbol{\varepsilon}}_{\mathbf{k}}$ is a polarization vector perpendicular to the direction $\mathbf{k}$ of propagation. Thus $\hat{\boldsymbol{\varepsilon}}_{\mathbf{k}}$ plays the role of $|sm\rangle$ in (62), yet it "knows" the momentum $\mathbf{k}$, as it is always perpendicular to it, whereas $|sm\rangle$ in (62) is independent of the momentum.

---

* Obviously, this argument does not apply to massless particles like the photon.

C. RELATIVISTIC QUANTUM THEORY

1. Field operators

Quantum field theory is the basic conceptual tool of particle physics. It is quite natural that this should be so, for particle physics is concerned with phenomena where the quantum of action and the finite speed of light both play an essential role. Already in classical physics the finite speed of signal propagation leads inexorably to the field concept—that is, to an infinite number of dynamical variables that extend throughout space-time. Hence one must expect quantum mechanical operators that are counterparts of classical fields to play a central role when the principle of relativity is imposed on quantum mechanics.

The objects considered to be particles in classical physics require a drastically different approach in relativistic quantum mechanics: they also acquire an indefinite number of degrees of freedom. Particles are produced by electromagnetic interactions, as in $\gamma A \to A e^+ e^-$, where A is some nucleus, and also in β-decay (see §D.3). A quantum theory of electrons, and other particles, must therefore be able to account for their appearance and disappearance, just as the quantum theory of electrodynamics must account for the creation and destruction of photons. In both cases, a physical process can cause a change in the number of degrees of freedom, and the state vectors of the system must therefore be essentially different from those that are used in nonrelativistic quantum mechanics.

The rather drastic measures that are called for if one is to transform quantum mechanics into a relativistic theory can be glimpsed from the following consideration. In nonrelativistic quantum mechanics, the time t is a parameter; it is not the eigenvalue of an observable. The coordinates x_i and momenta p_i of a particle are operators that satisfy the commutation rule

$$[x_i, p_j] = i\delta_{ij}. \tag{1}$$

From (1) it can be shown[*] that if x_i has the continuous spectrum $(-\infty, \infty)$, so does p_i. If we wish to amend quantum mechanics so as to conform to the principle of relativity, it is natural to generalize $\mathbf{r} = (x_1, x_2, x_3)$ to an operator that represents the position 4-vector, whose time component is the operator

---

[*] See, for example, Dirac (1958), § 23; and Gottfried (1966), § 29.

t_{op}, and **p** to the momentum 4-vector, whose time component is the Hamiltonian H. If (1) is to be a portion of a covariant set of laws, it would have to be complemented by the commutation rules

$$[t_{op}, H] = -i, \tag{2}$$

$$[x_i, H] = 0, \tag{3}$$

$$[t_{op}, p_i] = 0. \tag{4}$$

The sentence following (1) tells us that if the ordinary time t is to be the eigenvalue of t_{op}, with the continuous spectrum $(-\infty, \infty)$, then (2) will also require the energy spectrum to be continuous from $-\infty$ to ∞! Equation (3) says $\dot{x}_i = 0$, and that the energy and position can be measured simultaneously with arbitrary precision. Obviously this attempt at a relativistic version of quantum mechanics fails disastrously.

How is one then to treat the space and time coordinates on an equal footing? Given that one cannot promote the time to the status of an operator on a par with the coordinates, one can try another option: to demote the coordinates to the same status as time—to a set of parameters. The 4-vector* of parameters $x = (t, \mathbf{r})$ can then be used to label operators, just at t labels operators in the Heisenberg picture [§B.1(b)].

An operator A (x) parameterized by a space-time position is, by definition, a *field operator*. In quantum field theory, every point in the space-time continuum is endowed with such field operators. Whether or not a particular operator is dynamically significant in some space-time region is determined by the state of the system. As in ordinary quantum mechanics, the state in question is, at least in principle, determined by the manner in which the system was prepared—e.g., a 4 MeV antineutrino incident on hydrogen, in which case the state will evolve a component containing a positron and neutron (see §D.4).

2. The quantum theory of the electromagnetic field

The oldest, best understood, and most thoroughly tested quantum field theory is quantum electrodynamics (QED). It has two essential ingredients: a quantum theory of the electromagnetic field, and a relativistic quantum theory of the point sources of that field. This theory is among the most successful in the history of physics. Its predictions have been verified to a precision comparable to those of planetary dynamics.

The quantum theory of the electromagnetic field must contain Maxwell's theory, in the same sense as ordinary quantum mechanics reduces to Newtonian mechanics under appropriate circumstances. Another requirement follows from the usual $\Delta p \Delta q$ uncertainty relations satisfied by the sources of

---

* We use the notation $V = (V_0, V_1, V_2, V_3)$ for 4-vectors. The Lorentz invariant product of V and W is $V \cdot W = V_0 W_0 - \mathbf{V} \cdot \mathbf{W}$.

the field: the electromagnetic fields **E** and **B** emanating from the sources must have well-defined restrictions on their measurability, otherwise knowledge of these fields could be used to violate the lower bound on $\Delta p \Delta q$ of a particle. This shows that $\mathbf{E}(x)$ and $\mathbf{B}(x)$ must be noncommuting operators, and that their uncertainty products $\Delta E_i(x)\Delta B_j(x')$ must be such as to guarantee $\Delta p \Delta q \gtrsim 1$. It can be shown* that this requirement essentially determines the commutation rules satisfied by the operators **E** and **B**.

In order to guarantee the classical limit, the Hamiltonian of the quantum theory is taken to be the same function of the field operators **E** and **B** as the total energy in the classical theory:

$$H = \frac{1}{2} \int \{|\mathbf{E}(\mathbf{r},t)|^2 + |\mathbf{B}(\mathbf{r},t)|^2\} \, d^3r. \tag{5}$$

Consequently, the equations of motion satisfied by the quantum fields **E** and **B** are just Maxwell's equations.

Despite this formal similarity between the classical and quantum theories of the radiation field, the latter is a far richer construct, because in the quantum theory one must consider the state vectors in addition to the dynamical variables. As already mentioned in §B.2, any eigenstate of (5) is specified by the number of photons $n_{\mathbf{k}h}$ of momentum **k** and polarization h contained in it.** The energy eigenvalue is then given by Eq. B(13). The ground state is the *vacuum*, $|\Omega\rangle$, for which $n_{\mathbf{k}h} = 0$ for all **k** and h.

There is an analogy between **E** and **B** on one hand, and the amplitude q of a quantum oscillator on the other. The expectation value of q vanishes in the eigenstates of energy, as do the expectation values of **E** and **B** in the eigenstates of (5). An eigenstate of q has no definite energy; by the same token an eigenstate of **E** (or **B**) does not contain a definite number of photons! Thus quantum field theory accommodates both the continuum (**E** and **B**) and corpuscular (photon) descriptions of electromagnetic phenomena. These are complementary descriptions, just as position and momentum provide complementary descriptions of a point particle's motion.

To give the reader a more concrete feeling for the field operators, we present, without proof**, some of their matrix elements. Again we drawn on the analogy between the oscillator and the fields. The operator q has nonvanishing matrix elements $\langle n|q|n'\rangle$ only between states that differ by one unit of excitation, $n' = n \pm 1$. In an analogous fashion, the field operators only have matrix elements between stationary states that differ by one photon. The most important elements are the ones between the vacuum state $|\Omega\rangle$, and the one-photon state, which we designate by $|\mathbf{k}h\rangle$:

$$\langle \Omega|\mathbf{E}(\mathbf{r},t)|\mathbf{k}h\rangle = N\hat{\boldsymbol{\varepsilon}}_{\mathbf{k}h} \exp[i(\mathbf{k} \cdot \mathbf{r} - \omega t)], \tag{6}$$

---

* See Appendix IV (Vol. II) and Gottfried (1966), §§3 and 53.
** The statements made in this and the following paragraphs will be proven in Vol. II, §II.A.

$$\langle \Omega | \mathbf{B}(\mathbf{r},t) | \mathbf{k}h \rangle = iN(\mathbf{k} \times \hat{\boldsymbol{\varepsilon}}_{\mathbf{k}h}) \exp[i(\mathbf{k} \cdot \mathbf{r} - \omega t)], \tag{7}$$

where $\hat{\boldsymbol{\varepsilon}}_{\mathbf{k}h}$ is a unit polarization vector, and N is a normalization factor. Equations (6) and (7) show that these matrix elements have the form of classical electromagnetic waves. These equations imply that the operators $\mathbf{E}$ and $\mathbf{B}$ can *destroy* a photon. There are also matrix elements for the inverse process, wherein a photon is *created* out of the vacuum, as in

$$\langle \mathbf{k}h | \mathbf{E}(\mathbf{r},t) | \Omega \rangle = N\hat{\boldsymbol{\varepsilon}}_{\mathbf{k}h} \exp[-i(\mathbf{k} \cdot \mathbf{r} - \omega t)]. \tag{6'}$$

We infer that there are two independent polarization vectors, in correspondence with classical electrodynamics. If we choose these to be the right- and left-handed circular polarizations, the vectors are

$$\hat{\boldsymbol{\varepsilon}}_{\mathbf{k}\pm 1} = \frac{1}{\sqrt{2}} (\hat{\mathbf{x}} \pm i\hat{\mathbf{y}}), \tag{8}$$

where $\hat{\mathbf{x}}$ and $\hat{\mathbf{y}}$ are unit vectors in the plane perpendicular to the propagation direction $\mathbf{k}$. On comparing with Eq. B(22), we see that they transform under rotations about the axis $\mathbf{k}$ like states having the projection of angular momentum $\mathbf{J} \cdot \hat{\mathbf{k}}$ equal to 1 or -1. According to the definition of helicity, Eq. B(63), we see that* *the photon has helicity* $h = \pm 1$, *but never* $h = 0$. The absence of $h = 0$ is simply the familar fact that there are no longitudinally polarized electro-magnetic waves.

According to our definition of the spin as the maximum helicity (recall §B.7), *the photon is a spin one particle.* But 'the nonexistence of the $h = 0$ state means that the photon's intrinsic angular momentum is not a conventional angular momentum. This is not a paradox, since the massless photon can never be viewed in a rest frame. If one amends Maxwell's equations so that their quantized version has massive quanta, one finds that there is a $h = 0$ state, and then this "photon" has a conventional intrinsic angular momentum.**

Because there is no state with helicity zero, it is impossible to build a one-photon state with vanishing total angular momentum by combining one unit of orbital angular momentum with the photon's spin to give $j = 0$. Consequently, a spherically symmetric one-photon state does not exist—a fact already exploited in §B.6.

The multipole fields are one-photon states having a definite parity and total angular momentum j. Electric and magnetic multipoles of order 2^j are, respectively, symmetric or antisymmetric linear combinations of states with $h = +1$ or -1, and angular momentum j.

---

* Any linear combination of $h = 1$ and -1 is also allowed; in general such a combination is elliptically polarized.
** This remark plays an important role in the theory of weak interactions. See Chap. VI (Vol. II).

3. The Dirac theory of spin $\frac{1}{2}$ particles

The Dirac equation for the electron was intended to be a relativistic generalization of the Schrödinger equation—in short, a one-particle theory. At one time it was thought to "explain" the spin $s = \frac{1}{2}$ of the electron, but we now know that this is not so. Equations of the Dirac type can be constructed for any s. At this time we have no understanding of the remarkable fact that the fundamental fermions of particle physics (electrons, neutrinos, quarks, etc.) all have spin $\frac{1}{2}$.

By treating the Dirac equation as if it were a one-particle equation, one can account for a variety of important facts; the magnetic moment of the electron, and the fine structure splittings in the hydrogen spectrum, are the outstanding examples. But the theory, when so interpreted, soon runs into paradoxes. All stem from the fact that the equation for a free particle not only has solutions with energy $E = \sqrt{p^2 + m^2}$, as it should, but also with $E = -\sqrt{p^2 + m^2}$. If one discards the negative energy solutions, one does not have a complete set of stationary states, and that leads to violations of the causality principle (see §6). If one keeps the negative energy solutions, the hydrogen atom "ground state" is very unstable against decay into negative energy states.

The resolution of these paradoxes can be phrased in two equivalent ways. By taking advantage of the Pauli principle, one can assert that the vacuum is that state where all negative energy levels are singly occupied by a "sea" of electrons, just as the electrons in an insulator completely fill a conduction band.* Excited states are then formed by placing electrons in positive energy states, and/or lifting electrons from negative to positive energy states. In the latter case one leaves a "hole," which behaves just like an electron except that it has the opposite charge. This is *the positron*, designated by e^+.

The second, equivalent, interpretation of the Dirac equation asserts that the solution $\psi(\mathbf{r}t)$ of the equation is not a number, but a field operator like **E** and **B**. Once again, matrix elements of ψ with the time-dependence $\exp(-i|E|t)$ represent the destruction of a particle with energy $E > 0$, as in Eq. (6), whereas matrix elements having the time-dependence $\exp(i|E|t)$ describe the creation of a particle with energy $E > 0$, in analogy with (6'). But in this case there is a fundamentally new feature, because the electron carries charge, whereas the photon does not, so if the word "particle" in the foregoing sentence were to refer to the electron in *both* instances, the operator ψ, when acting on a state of definite charge Q, would produce a superposition of states with charges $Q + e$ and $Q - e$. As we shall learn in §II.A (Vol. II) and Appendix IV, the principles of charge conservation and relativistic invariance lead to the requirement that any field operator, such as ψ, must produce a state of well-defined charge when acting on any such state. Consequently the particles that are destroyed and created by ψ cannot be

---

* The total charge of the negative energy "sea" is disregarded in this description, whereas in insulators the charge of the electrons is compensated by the positive ions.

identical; indeed, *they must have equal but opposite charge. Hence the existence of the electron implies the existence of the positron.*

To summarize, the solution ψ of the Dirac equation, when interpreted as a field operator, destroys electrons or creates positrons; its Hermitian adjoint $\psi^\dagger$ creates electrons or destroys positrons. With this interpretation the theory can, in a consistent manner, account for pair creation and destruction without invoking the concept of a negative energy sea. If the Dirac field is in the presence of a fixed charge $+e$, the matrix element of $\psi(\mathbf{r}t)$ between the vacuum and one-electron states defines a "wave function" that provides a relativistic description of the hydrogen spectrum.

The prediction of the existence of the positron and, quite generally of antimatter, was a historic triumph. The symmetry between matter and antimatter that it revealed is one of the constant themes of particle physics.

In any event, whether one uses the language of "hole" theory, or of field theory, it is clear that the Dirac equation describes a system with an infinite number of degrees of freedom. That is to be expected from the argument of §1.

4. Interactions

We have briefly discussed free photons, electrons, and positrons, but all observable phenomena are due to interactions. In quantum electrodynamics (QED) the interaction is a term H_{em} in the Hamiltonian, adapted from classical electrodynamics, which involves the electromagnetic and Dirac field operators:

$$H_{em} = \int \mathbf{j} \cdot \mathbf{A} \, d^3x, \tag{9}$$

where $\mathbf{A}$ is the vector potential (i.e., $\mathbf{B} = \boldsymbol{\nabla} \times \mathbf{A}$), and $\mathbf{j}$ is the current due to electrons and positrons. This current is a bilinear form in the electron–positron field operators ψ and $\psi^\dagger$ [see §II.A.3 (Vol. II)]. The operator $\mathbf{A}$ creates and destroys photons, just as the field-strength operators $\mathbf{E}$ and $\mathbf{B}$ do; the bilinear form $\psi^\dagger \psi$ either changes the state of an electron or positron (when both $\psi^\dagger$ and ψ act on electrons or positrons, respectively), or produces or annihilates an electron–positron pair (when one factor in $\mathbf{j}$ acts on electrons, while the other acts on positrons). Thus H_{em} has matrix elements that allow the following basic transitions:

$$e^- \leftrightarrow e^- \gamma \tag{10a}$$

$$e^+ \leftrightarrow e^+ \gamma \tag{10b}$$

$$e^+ e^- \leftrightarrow \gamma \tag{11}$$

Because H_{em} is Hermitian, the amplitude for any transition $a \to b$ is the complex conjugate of that for $a \leftarrow b$. Symmetry between electrons and positrons requires that the matrix elements for (10a) and (10b) have the same absolute value.

The translation invariance of H_{em} implies that the total linear momenta on both sides of (10) or (11) are equal. There is no such condition on the energy. Indeed, were one to demand energy *and* momentum conservation, one would find that all the processes in (10) and (11) are forbidden. This must be so, otherwise free electrons would radiate, free photons would materialize into pairs, etc. In short, (10) and (11) are only ingredients, or subprocesses, in any actual process, as we shall soon see.

An intuitively appealing depiction of the manner in which the subprocesses (10) and (11) combine to form an actual process is provided by *Feynman diagrams*. How the diagrams emerge from the source-field interaction operator H_{em}, and to what extent one can ascribe a physical reality to the diagrams, will be discussed in §II.B (Vol. II). But even before one understands all that, one should not be afraid to use the diagrams as a conceptual aid in visualizing and analyzing a process.

Feynman diagrams have two types of building blocks: *vertices* and *propagators*. A vertex describes a fundamental transition due to the interaction in question. In the case of QED, these transitions are given by (10) and (11), and are represented by the vertices

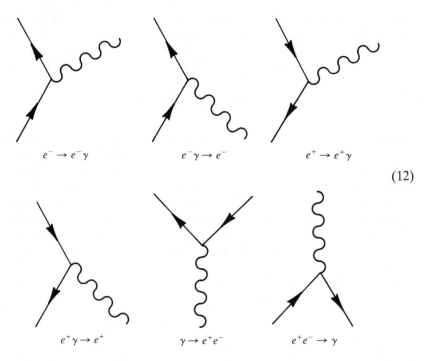

$$e^- \to e^- \gamma \qquad e^- \gamma \to e^- \qquad e^+ \to e^+ \gamma$$

$$e^+ \gamma \to e^+ \qquad \gamma \to e^+ e^- \qquad e^+ e^- \to \gamma$$

$$\tag{12}$$

Here straight lines with arrows represent electrons or positrons, and wavy lines photons.

One can think of these and other Feynman diagrams as representing the temporal evolution of the process. In (12) we have used the convention that time increases as one looks toward the top of the page; sometimes (as in Chap. III, Vol. II) we shall find it more convenient to draw the diagrams as if time increases from left to right.

Note that positrons are only distinguished from electrons by the direction of their arrow: a positron in the *final* state (as in $\gamma \rightarrow e^+e^-$) is depicted as if it were *entering* the interaction, whereas a positron in the initial state (see $e^+e^- \rightarrow \gamma$) looks as if it were leaving the interaction! This convention for drawing Feynman diagrams aptly captures the complete symmetry between e^+ and e^-, as well as the requirement that charge be conserved in each and every interaction. Furthermore, the convention emphasizes that all the processes in (12) are related to each other by *crossing*, a concept that will be introduced in the next section. For these reasons the same convention will be used for all the antiparticles that we shall encounter.

As we have already said, actual processes involve two or more of the basic transitions (10–11), and can, for that reason, be depicted by diagrams obtained by joining two or more vertices from (12). Thus a Feynman diagram contributing to Compton scattering is*

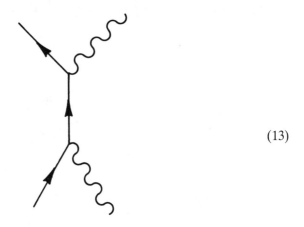

$$(13)$$

Here the intermediate state in $e^-\gamma \rightarrow e^- \rightarrow e^-\gamma$ is represented by an electron line, or electron propagator, depicting the propagation of the electron between the absorption and emission acts. The total energies E of the initial and final states are equal, but the conservation law does *not* imply that the electron in the intermediate state has this values of E. This is best seen in the c.o.m. (center-of-mass) frame, because there $\mathbf{P} \equiv 0$, whereas $E > m$, and therefore $E \neq \sqrt{P^2 + m^2}$. There is no inconsistency, however, because the energy of an intermediate state has an uncertainty of order $(\Delta t)^{-1}$, where Δt is the duration of the collision, which depends on the detailed dynamics of

---

* This is not the only diagram for Compton scattering; see §II.B.3 (Vol. II).

the process in question. On the other hand, the total momentum is conserved at each vertex; it is the same in the initial, intermediate, and final states.

As a second example of an electrodynamic process, consider pair annihilation. While the process $e^+e^- \rightarrow \gamma$ cannot actually occur, pair annihilation into two photons, as the two-step chain $e^+e^- \rightarrow \gamma e^+e^- \rightarrow \gamma\gamma$, is allowed by energy and momentum conservation. The corresponding diagram is

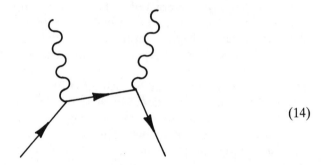

$$(14)$$

A third example is afforded by elastic e^+e^- scattering, for which the basic diagrams are

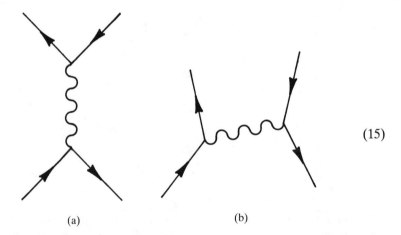

$$(15)$$

(a)   (b)

Here we have a photon in the intermediate state, depicted by a photon line, or photon propagator. It represents the electromagnetic field that propagates from the annihilation to the creation point in (a), and between the electron and positron in (b). In any process that actually occurs, such as those depicted in (13–15), the total energy and momentum of the initial state must equal those of the final state, but as already explained, energy conservation does not determine the energy of the intermediate states. The structure and evolution of the intermediate state is represented by one or more internal lines (or propagators) in the Feynman diagram.

Because the transitions (12) all have equal amplitudes, sets of processes in QED are related to each other. For example, $e^- \gamma \to e^- \gamma$, $e^+ \gamma \to e^+ \gamma$, $e^+ e^- \to \gamma\gamma$ and $\gamma\gamma \to e^+ e^-$ all have essentially the same amplitude. This is also brought out by the Feynman diagrams; for example, if one looks at (14) from left to right, it becomes a Compton amplitude. The cross sections for these related processes are not equal, however, because the cross section involves more than the transition amplitude: kinematical factors describing the flux of incident particles, final and initial state wave functions, etc. These change from process to process.

5. Crossing and charge conjugation

The kinship between Compton scattering and pair annihilation is a general feature of relativistic quantum theory. Let $a, b, \ldots$ be particles, $\bar{a}, \bar{b} \ldots$ their respective antiparticles, and assume that they can participate in the reaction

$$a + b \to c + d. \tag{16}$$

Next we define the operation of *crossing*, which consists of taking a particle from one side of (16) to the other, while changing it from particle to antiparticle. All reactions obtained from (16) by crossing one or more particles are then also physically possible processes, provided that energy and momentum are conserved. Thus from (16) we infer that

$$a + \bar{c} \to \bar{b} + d,$$
$$\bar{c} + \bar{d} \to \bar{a} + \bar{b}, \tag{17}$$

etc., are possible processes related to (16). If one of the particles, say c, is heavy enough, (16) also implies the existence of the decay*

$$c \to a + b + \bar{d} \tag{18}$$

We have not yet defined what, in general, we mean by antiparticle. Let a be any particle. If a has no attributes beyond linear and angular momentum (which include energy and spin), then a is its own antiparticle. The photon and the neutral pion are such particles. If the particle a has any attributes beyond momentum and spin (such as charge Q) the antiparticle $\bar{a}$ has the opposite attributes (such as charge $-Q$). The neutron (n) and proton (p) have such attributes. In the case of p, the charge obviously distinguishes p from $\bar{p}$, while n, though electrically neutral, has a magnetic moment that has

---

* Naturally two-body decay $c \to a + b$ may also be possible if $m_c > m_a + m_b$. There are many examples of this: all radiative transitions ($a \to b + \gamma$), or the decay of the pion.

the opposite sign from that of $\bar{n}$.* The case of the neutrino is more subtle. As we shall learn in §D.4, it is an experimental fact that neutrinos are not identical to their antiparticles.

The replacement of particles by their antiparticles is called the operation of *charge-conjugation*, or C for short. This is something of a misnomer, because it is also used for particles, like the neutron, which have no charge.

6. Causality and its consequences

In §2 we mentioned that the electric and magnetic field operators must satisy certain commutation rules so as to guarantee $\Delta p \Delta q \gtrsim 1$. These commutation rules must also satisfy a condition imposed by causality. This requirement is totally general, and is not restricted to electrodynamics. In view of the important consequences that flow from it, we shall state the condition, and summarize the consequences. A more detailed discussion will be found in Appendix IV (Vol. II).

We recall that when two observables A and B can be measured simultaneously, A and B commute with each other, and their uncertainty product has zero as a lower bound: $\Delta A \Delta B \geq 0$. Now let $A = A(x)$ be some field operator at the space-time point x, and $B(x')$ any other field operator at another point x'. If x and x' cannot be connected by a light signal, i.e., if $(\mathbf{r} - \mathbf{r}')^2 \geq (t - t')^2$, one must be able to measure $A(x)$ and $B(x')$ simultaneously to arbitrary precision. Therefore

$$\Delta A(\mathbf{r},t) \Delta B(\mathbf{r}',t') \geq 0 \quad \text{if } (\mathbf{r} - \mathbf{r}')^2 > (t - t')^2. \tag{19}$$

This is called *the causality condition*.

Remarkably enough, a number of important conclusions can be derived from (19) without knowing the structure of the Hamiltonian.** We shall list some of these, and discuss them briefly:

1. The connection between spin and statistics: half-integer spin particles obey Fermi–Dirac statistics, while integer spin particles obey Bose–Einstein statistics. This is an empirical fact for electrons, protons, neutrons, photons, and pions which cannot be explained by non-relativistic quantum mechanics.

2. Particles have the opposite intrinsic parity to their own antiparticles if they are fermions, and the same intrinsic parity if they are bosons. For example an e^+e^- pair in an S-state has odd parity whereas a $\pi^+\pi^-$ pair in an S-state has even parity, because the pion is a spin-zero particle. This theorem plays an important role in the spectroscopy of positronium (e^+e^- bound states), and in meson spectroscopy.

---

* In §D we shall also define the attribute "baryon number"; possessed by neutrons and protons, which changes sign in going from particles to antiparticles.

** Certain other highly plausible assumptions must be made, such as Lorentz covariance, a lower bound to the energy spectrum, etc.

3. Particles and their antiparticles have the same mass, equal but opposite charges and magnetic moments, and if they are unstable, the same lifetime, even if the Hamiltonian is *not* invariant under the interchange of particles and antiparticles! This is called the CPT theorem, and it is of considerable importance, for it is known (see §E.9) that charge-conjugation is not an exact symmetry. This theorem only applies to the lifetime, i.e., to the total decay rate. If there are several decay modes, the branching ratios need not be the same for particle and antiparticle.

4. Let $\sigma_{tot}(ab)$ be the total cross section for elastic and inelastic collisions of any two particles a and b, and $\sigma_{tot}(a\bar{b})$ the corresponding cross section for $a\bar{b}$ collisions; then $\sigma_{tot}(ab) - \sigma_{tot}(a\bar{b})$ tends to zero in the high energy limit. This is known as the Pomeranchuk theorem.

To within the present experimental accuracy, the masses and lifetimes of all particles are equal to those of their antiparticles, in conformity with the CPT theorem. We shall quote just a few facts attesting to these symmetries between particles and antiparticles. If $\tau^{\pm}$ are the lifetimes of the positively and negatively charged pions, then

$$\frac{\tau^{+} - \tau^{-}}{\tau^{+} + \tau^{-}} = (1.0 \pm 1.4) \times 10^{-3}.$$

The ratio of the electron and positron masses is one to a part in 10^7, and their magnetic moments have a ratio that is -1 to a part in 10^{11} (Schwinberg, 1981). At the highest energies presently available, the total cross sections for pp, $\pi^+ p$, and Kp collisions are all approaching the cross sections for $p\bar{p}$, $\pi^- p$, and $\bar{K}p$, respectively, in conformity with the prediction of the Pomeranchuk theorem (see Perkins, 1982, Fig. 4.31).

D. NUCLEAR PHENOMENA

1. The nuclear spectrum

The theory of atomic and molecular phenomena deals with the interactions of electrons and nuclei. Both constituents are treated as particles with fixed properties, such as charge, mass, spin, and magnetic moment. The relevant energy exchanges in this set of phenomena are in general well below 10^5 eV, that is, well below the threshold for nuclear excitation. Hence in atomic and molecular phenomena, nuclei can be considered as "elementary" particles along with the electrons.

A different realm of phenomena appears when the energy exchanges are in the range of 10^5–10^7 eV: atomic nuclei cease to be "elementary." They exhibit a spectrum of excited states, and it is possible to take nuclei apart by removing some of the constituents. The discovery of the neutron led to the recognition that nuclei are composed of protons and neutrons, both called *nucleons*. The proton carries the opposite charge to the electron, the neutron is neutral. The mass of the proton is $m_p = 938$ MeV $= 1836m_e$, and the mass of the neutron is a tenth of a percent higher: $m_n - m_p = 1.29$ MeV $= 2.52m_e$. The nucleons are particles with spin one-half. Both carry a magnetic moment different from what one would expect from the Dirac equation (see §II.A, Vol. II), which does account correctly for the electron's magnetic moment. According to that equation the proton's moment μ_p should be one proton Bohr magneton, $\mu_B = e\hbar/2m_pc$ (or $e/2m_p$ in our units), and the neutron, being neutral, should have no moment. Experiments show that $\mu_p \approx 2.6\mu_B$ and $\mu_n \approx -1.9\mu_B$! This is a first indication that the nucleons are not simple elementary point particles. More striking evidence is provided by electron-proton scattering, which shows (see §III.A, Vol. II) that the proton's charge is spread over a length of about one fermi (10^{-13} cm $= 1$ fm).

Since the neutron is uncharged, and the protons repel each other because of their charge, there must exist a special interaction, the *nuclear force*, which keeps the nucleons together within the nucleus. Scattering experiments show that this is a strong force with a finite range of about 2 fm, mostly attractive but turning repulsive at distances smaller than about 0.7 fm. By "strong" we mean, for example, that at a separation of a fermi the nuclear attraction between two protons is about 10 times larger than their Coulomb repulsion.

The term "finite range" refers to the fact that the force goes to zero exponentially, like $e^{-r/b}/r$, in a characteristic distance b (see Fig. 2). Its strength depends to some extent on the relative spin orientation or, equivalently, on the symmetry of the state under interchange of the two interacting nucleons; recall Eqs. B(45) and (46). This force between nucleons is a manifestation of the so-called "strong interaction" which plays a prominent role in the subnuclear realm.

Once the force between nucleons is determined and expressed in terms of a potential energy, quantum mechanics can be used to predict the spectra and other properties of nuclei, and the interaction between nucleons and nuclei. It turns out that the nuclear force is not strong enough to require a relativistic treatment of the dynamics within the nucleus. Nonrelativistic quantum mechanics suffices for an approximate description, and is very successful in explaining the properties of nuclei and of nuclear reactions.

Nuclei possess spectra of bound states that have a considerable similarity to atomic spectra. Each nuclear state possesses an angular momentum quantum number J. As in atomic dynamics, this is a natural consequence of the isotropy of space. J is an integer or half-integer, depending on whether the number of constituent nucleons is even or odd, because nucleons have an intrinsic spin of $\frac{1}{2}$. The most striking distinction between nuclear and atomic spectra is the typical magnitude of the energy differences. A simple estimate of that magnitude can be made by the following crude method. The main attractive part of the nuclear force (Fig. 2) can be replaced by a Coulomb

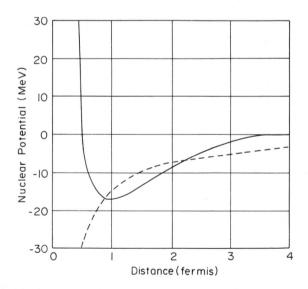

FIG. 2. Sketch of the potential of the nuclear force as a function of internucleon separation as measured in fermis, 10^{-13} cm. (This curve is not quantitative, because it ignores the dependence on spin and isospin, or equivalently, symmetry.) For comparison, the dashed curve gives the attraction of two opposite, but equal, charges $3.7e$.

attraction α_N/r, where $\alpha_N \sim \frac{1}{10}$. This rough approximation does not take into account the repulsion at small distances, nor the exponential decrease at large distances, but says that the force is roughly like the attraction of two opposite charges of $3.7e$. On replacing the fine structure constant α and the electron mass m_e by α_N and the nucleon mass m_N in the expressions for the Rydberg and the Bohr radius [see Eqs. B (57) and (58)], we get the nuclear Rydberg and the nuclear Bohr radius a_N as a crude estimate of the order of magnitude of nuclear excitation energies and dimensions:

$$\mathrm{Ry}_N = \frac{1}{2} m_N \alpha_N^2 \sim 5 \text{ MeV}, \qquad a_N = \frac{1}{m_N \alpha_N} \sim 2 \text{ fm.} \qquad (1)$$

Consequently nuclear excitation energies are of the order of a few MeV, and nuclear dimensions are of the order of a few fermis. It must be emphasized that such estimates can only serve as a first orientation. After all, the nuclear force is repulsive at small distances and the attractive part of the nuclear potential decreases exponentially. This is why the deuteron has no bound states with nonzero orbital angular momentum; the centrifugal effect drives the particles away from each other to distances where the nuclear force is no longer appreciable. Nuclei with more than two constituents do possess bound states with nonzero orbital angular momenta; in this case one nucleon acts on the next, and this in turn acts on its neighbors, so that the effective binding has larger range.

As we have said, one can attain a rather sound understanding of nuclear physics by applying nonrelativistic quantum mechanics to a system of* B structureless nucleons interacting through a force like the one of Fig. 2. But it must also be said that nuclear physics is not on as firm foundations as atomic physics. As we shall see, nucleons are themselves complex objects with internal excitations. Consequently, the interactions between them can only be approximated by the notion of a potential. Furthermore, internal nuclear motions can, on occasion, excite the nucleons. This means that a nucleus of atomic number B is only a system of B nucleons at a certain level of accuracy and sophistication. At a deeper level, there is no well-defined boundary between nuclear and subnuclear physics.

2. Isotopic spin

A new quantum number appears in nuclear spectra which has great significance in particle physics: the "isotopic spin" (isospin). Its importance is based on the empirical fact that the proton and the neutron are subject to the same nuclear forces, and that a proton can transform itself into a neutron, or vice versa, in processes due to the weak interaction, which will be treated

---

* We use the symbol B here, because this turns out to be the quantity called baryon number in subnuclear physics. In nuclear and atomic physics it is conventional to designate this number by A.

in the next section. It is therefore useful to consider the proton and the neutron as two states of one particle—the nucleon. The two states differ in their electrical properties, but are equivalent insofar as nuclear forces are concerned.

Thus the nucleon is a two-level system in the sense of §B.3(e). The proton and the neutron states $|p\rangle$ and $|n\rangle$ [called $|\alpha\rangle$ and $|\beta\rangle$ in §B.3(e)] span an abstract space $\mathscr{C}_2^I$ that has the same properties as the space $\mathscr{C}_2$ of the spin $\frac{1}{2}$ doublet. This allows us to introduce the concept of *isotopic spin*, or isospin for short. By definition, the proton and neutron form a doublet with total isospin $I = \frac{1}{2}$, and "projections" $I_3 = \pm\frac{1}{2}$, respectively.

As in Eq. B(31), the proton and neutron can be written as basis spinors in $\mathscr{C}_2^I$:

$$|p\rangle = \begin{pmatrix} 1 \\ 0 \end{pmatrix} \qquad |n\rangle = \begin{pmatrix} 0 \\ 1 \end{pmatrix}. \tag{2}$$

The analogues of the spin $\frac{1}{2}$ angular momentum matrices are the three 2×2 matrices $\frac{1}{2}\tau_a$, where $\tau_1 = \sigma_x$, $\tau_2 = \sigma_y$, and $\tau_3 = \sigma_z$ are the three Pauli matrices [Eq. B(26)], but they are represented by a distinct symbol because they operate in a different space.

The nucleon has both ordinary spin and isospin, and in (2) the former has been suppressed as we now focus on the latter. A complete nucleon wave function therefore has two two-valued quantum numbers, one specifying its spin state, the other its isospin. The σ_i act on the former, the τ_a on the latter. Whenever we use the term spin, as opposed to isospin, we always refer to the intrinsic angular momentum.

The existence of three "rotation" generators $\frac{1}{2}\tau_a$ allows us to introduce an *abstract Euclidean 3-dimensional isospin space* $\mathscr{C}_3^I$; the relationship between $\mathscr{C}_3^I$ and $\mathscr{C}_2^I$ is precisely the one that exists between ordinary Euclidean 3-space $\mathscr{C}_3$, and the space $\mathscr{C}_2$ of the $j = \frac{1}{2}$ doublet [recall §B.2 (e)]. In particular, the Pauli matrices combine to form

$$\vec{\tau} = (\tau_1, \tau_2, \tau_3), \tag{3}$$

which is *a vector in isospin space*.* Throughout this book we shall use the notation $\vec{V}$ to designate vectors in abstract spaces that refer to quantum numbers unrelated to the ordinary 3-space $\mathscr{C}_3$.

According to (2), the proton and neutron are the "up" and "down" numbers of the nucleon doublet, where up and down refer to the 3-direction in $\mathscr{C}_3^I$ (*not* in ordinary space, of course). If we "rotate" through 180° about the 2-axis of $\mathscr{C}_3^I$, $|p\rangle$ is "turned" into $|n\rangle$, and vice versa. An arbitrary transformation, like Eq. B(40), would transform $|p\rangle$ into some linear combination of $|p\rangle$ and $|n\rangle$. In contrast to an arbitrary spinor, such an arbitrary isospinor does

---

* To be precise, if U is given by Eq. B(40), with σ replaced by $\vec{\tau}$, $U\vec{\tau}U^{\dagger}$ is just the combination of τ_a's that we would get if we rotated a 3-vector by $(\hat{n},\theta)$.

not represent a state that can actually be observed, because the electromagnetic interaction always distinguishes p from n, and therefore selects a direction in isospin space. We have, by convention, called this preferred axis the 3-direction. The analogous situation would occur with ordinary spin if space were pervaded by a magnetic field **B**, which would then produce a Zeeman splitting of all spin doublets, and select a preferred direction. On the other hand, if **B** were weak enough so that the Zeeman splittings were very small compared to typical level spacings, we could, to a good approximation, ignore them, and then all linear combinations of spin up and down with respect to **B** would be degenerate and on an equal footing. It is this last circumstance which is analogous to the nuclear situation, because the strong interaction, which is responsible for nuclear forces, appears not to single out any preferred direction in isospin space, while the electromagnetic interaction plays a very subsidiary role in most nuclear phenomena. *To the extent that one ignores electromagnetic effects, all directions in isospin space are therefore equivalent.* The evidence for this assertion will be presented below.

From the τ_a we can form several operators that play an important role in nuclear physics. First we define the isospin components,

$$I_a = \tfrac{1}{2}\tau_a. \tag{4}$$

The isovector $\vec{I} = (I_1, I_2, I_3)$ is the analogue of the angular momentum **J**, and is called the *isospin* (operator). It has the same commutation rules as angular momentum. The electric charge is

$$Q = I_3 + \tfrac{1}{2}, \tag{5}$$

since $I_3 = \tfrac{1}{2}$ and $-\tfrac{1}{2}$ for p and n. The operators

$$I_\pm = I_1 \pm iI_2 \tag{6}$$

will be important in β-decay, because $I_+|n\rangle = |p\rangle$, $I_-|p\rangle = |n\rangle$ [see the analogue, Eqs. B(34) and (35)].

For systems of B nucleons we define the *total isospin* by the analogue of Eq. B(41):

$$\vec{I} = \sum_{n=1}^{B} \vec{I}^{(n)} \tag{7}$$

where $\vec{I}^{(n)}$ is the operator (4) for the nth nucleon. We can then construct simultaneous eigenstates of $(\vec{I})^2$, with eigenvalue $I(I + 1)$, and of I_3.

Consider the two-nucleon system, and its interactions. Each nucleon is described by an isospinor like (2). Because of the complete analogy with angular momentum we conclude from Eqs. B(45) and (46) that the eigenstates of definite (I, I_3) are

	$I_3 = 1$	$I_3 = 0$	$I_3 = -1$
$I = 1$	pp	$\dfrac{1}{\sqrt{2}}(pn + np)$	nn
$I = 0$		$\dfrac{1}{\sqrt{2}}(pn - np)$	

$$(8)$$

The triplet state is symmetric, the singlet state is antisymmetric, when the nucleons are interchanged. Under rotations in the isospin space $\mathscr{E}_3^I$, the triplet transforms like a vector, while the singlet is an invariant.

We now discuss the invariance of the nuclear Hamiltonian H when protons and neutrons are interchanged. Invariance under such an interchange does not require invariance under all rotations in the isospin space $\mathscr{E}_3^I$. For example, if H contains a term CI_3^2, this has the same value $\frac{1}{4}C$ for the one-nucleon states* $|p\rangle$ and $|n\rangle$, but it is only invariant under arbitrary rotations about the 3-axis of $\mathscr{E}_3^I$. For the two-nucleon system, this term has the eigenvalue C for $|pp\rangle$ and $|nn\rangle$, and zero for $|pn\rangle$ and $|np\rangle$. This reveals the lack of full symmetry, because the three $I = 1$ states are not degenerate if $C \neq 0$, as they must be if H is an invariant in $\mathscr{E}_3^I$. It is, however, an empirical fact that the energies of two-nucleon states depend only on whether they are symmetric or antisymmetric under interchange (i.e., $I = 1$ or $I = 0$), and not on whether the state contains two "equal" or two "unequal" nucleons. *Thus nature has chosen invariance under all rotations in isospin space as the realization of the equivalence of neutron and proton.*

In analogy to Eq. B(48) we can ascribe the difference in energy between the $I = 1$ and the $I = 0$ states to a term in the Hamiltonian proportional to $(\vec{\tau}^{(1)} \cdot \vec{\tau}^{(2)})$, where the vector operators $\vec{\tau}^{(i)}$ have been defined previously. The product $(\vec{\tau}^{(1)} \cdot \vec{\tau}^{(2)})$ is invariant under rotations in $\mathscr{E}_3^I$. Its eigenvalue is 1 for $I = 1$, and -3 for $I = 0$, in analogy with Eq. B(49); it splits the symmetric triplet from the antisymmetric singlet.

The Pauli principle acquires a more general significance when isospin is introduced. Now the proton and the neutron are no longer different particles, but the *same* particle (the nucleon) with different eigenvalues of a new intrinsic coordinate. Hence a state consisting of several nucleons must be antisymmetric under exchange of any two nucleons, and not only under the exchange of two neutrons or protons. For example, when two nucleons are in the symmetric $I = 1$ state, they must be antisymmetric either in the spin coordinates, or in the space coordinates; if they are in the antisymmetric

---

* The expression CI_3^2, though not fully invariant in $\mathscr{E}_3^I$, does not distinguish between p and n, because it is also invariant under a rotation through 180° about the 1- (or 2-) axis, which interchanges n and p.

$I = 0$ state, the remainder of the state must be symmetric, i.e., either symmetric in both space and spin, or antisymmetric in both.

The simplest manifestation of the $I = 0$ state in the two-nucleon system is the deuteron. The $I = 1$ two-nucleon system has no bound states, only scattering states, and therefore the ground state is an $I = 0$ state. Because of the Pauli principle, that state must be symmetric in the other coordinates. Since it is an orbital S-state (symmetric), as most ground states are, it must also be symmetric in the spins. Hence the deuteron has a total spin of 1 in its ground state.

The three nuclei ^{12}B, ^{12}C, ^{12}N consist of five protons and five neutrons each, plus a pair of valence nucleons, pp in ^{12}N, pn in ^{12}C, nn in ^{12}B. The five protons and five neutrons are in an $I = 0$ state, and form an inert core. We concentrate our attention on the valene pair, which provides an example of two-nucleon states. The pair is in a triplet state ($I = 1$), or a singlet state ($I = 0$), of different energy. In most nuclei with an equal number of protons and neutrons the lowest states are isospin singlet states.* The substates of the triplet give rise to ^{12}B, ^{12}C, and ^{12}N, whereas the singlet (pn only!) can only be ^{12}C. We therefore expect all three nuclei to have degenerate $I = 1$ states, whereas ^{12}C should have additional states with $I = 0$. Actually the degeneracy of the isospin substates is only approximate, since the p–n mass difference, and the electrostatic repulsion, introduce small energy differences. In Fig. 3 we see that the lowest states of ^{12}B and ^{12}N have a spectrum similar to the $I = 1$ spectrum of ^{12}C. The $I = 1$ states in ^{12}C lie midway between those of ^{12}B and ^{12}C. The $I = 0$ states appear only in ^{12}C.

In general, a state consisting of Q protons** and N neutrons possesses the quantum number $I_3 = \frac{1}{2}(Q - N)$. It may belong to any multiplet with isospin $I = |I_3|, |I_3| + 1$, etc., but I can never be larger than $\frac{1}{2}(N + Q)$. All substates of a given I have approximately the same nuclear energy. As in the two-nucleon case, the substates of a given I have the same symmetry with respect to exchanges of any nucleon pair, since the operators $I_\pm$, which transform one into the other [(see Eqs. (6) and (7)] are symmetric under such exchanges, and cannot change the symmetry of the states.

The n–p mass difference and Coulomb repulsion do not only spoil the symmetry of the Hamiltonian; they also effect the wave functions, which then differ somewhat as I_3 changes within an I-multiplet.

3. Electromagnetic and weak transitions between nuclear states. Neutrinos

As our preceding discussion has indicated, all the bound states of B nucleons should be viewed as being the levels of one system. Let us designate such a

---

* This comes from the Pauli principle: the antisymmetry in isospin requires symmetry in the other coordinates. The latter circumstance usually lowers the energy.

**We designate the charge by Qe, because that is the customary notation in subnuclear physics. In atomic and nuclear physics the charge is conventionally denoted by Ze.

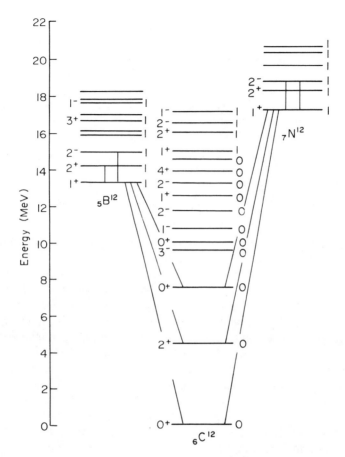

FIG. 3. The level spectrum of the 12-nucleon system. The numbers on the left of each level give the angular momentum J and parity, those on the right give the isospin. The connecting lines are the most important transitions; the vertical ones are electromagnetic, the skewed ones are lepton pair emissions $(e^-\bar{\nu}$, or $e^+\nu)$.

level by $\Phi_n(Q)$, where Q is the atomic number (the number of protons), and n specifies other quantum numbers. We do not display B as it is common to all these levels.

Transitions within one nuclear family occur by the emission or absorption of *two* types of "radiation." Just as in atoms, transitions

$$\Phi_n(Q) \leftrightarrow \Phi_{n'}(Q) + \gamma \tag{9}$$

between levels of *the same charge* can occur by the emission or absorption of photons. The same selection rules restrict the changes of angular momentum and parity, and determine the multipole type of the emitted radiation. In nuclear transitions, higher multipole radiations are not as hindered as in atomic transitions because the ratio R/λ (size to wavelength) is about 10 times larger, i.e., of order α_N instead of α (recall the discussion on p. 31).

Nuclear physics reveals another important transition mode, in which the nuclear charge changes by $\pm e$, i.e., $\Delta Q = \pm 1$. Nuclear levels that perform such transitions are called β-emitters, and the process in question is called β-decay. The transition $\Phi_n(Q) \to \Phi_{n'}(Q \pm 1)$ is accompanied by the emission of a pair of particles. On of these is an electron or positron, depending on whether $\Delta Q = +1$ or -1. The second particle is a neutrino. Schematically, we have

$$\Phi_n(Q) \to \Phi_{n'}(Q + 1) + e^- + \bar{\nu}, \tag{10^-}$$

$$\Phi_n(Q) \to \Phi_{n''}(Q - 1) + e^+ + \nu. \tag{10^+}$$

The evidence that the neutrinos in e^- and e^+ emission are distinct, and each other's antiparticles, will be discussed presently.

That at least two particles must be emitted follows from angular momentum conservation. As all the $\Phi_n(Q)$ consist of B nucleons, they have an integer or half-integer total angular momentum J depending only on whether B is even or odd. Consequently, any transition within the nuclear family $\{\Phi_n(Q)\}$ involves an integer change of J, and must be accompanied by the emission of an even number of fermions. Furthermore, if only one particle were emitted, its energy would be given uniquely by the change of nuclear mass. The $e^{\mp}$ energy spectrum is actually continuous, indicating that the available energy is shared with other decay products. The two-body decays (10) are therefore the simplest possibility, once one knows that an e^- or e^+ is radiated. This argument shows that the neutrino is a fermion. Detailed studies of the angular and energy distributions in β-decay show that the neutrino spin is $\frac{1}{2}$, and that its mass is less than $10^{-4} m_e$. Hence the neutrino has a mass that is far smaller than all the masses we have encountered. It is entirely possible that $m_\nu \equiv 0$, but no firmly established principle appears to require this.

β-decay is a process that is very slow compared to other nuclear processes; in the fastest instances the lifetime is of order milliseconds, while in the slowest cases the lifetime is greater than the age of the universe. The interaction causing this emission is therefore called the weak interaction.

Electrons and neutrinos belong to the so-called lepton family of particles. The word "lepton" was taken from the Greek word for "light" at a time when the known leptons included the electron, the neutrino, the muon μ, and a second type of neutrino associated with it.* Today we know that "lightness" is not a universal characteristic of leptons, for there exists a charged lepton τ which has about twice the mass of the nucleon. The μ, τ, and their neutrinos will be discussed in §E.9.

---

* Because the electron and muon neutrinos are distinct, we should have written ν_e in (10), but as nuclear transitions only involve ν_e, we have dropped the subscript "e" here.

The creation of a lepton pair in a transition caused by the weak interaction is somewhat analogous to the emission of a photon by the electromagnetic interaction. Both the lepton pair, and the photon, are "created" when the transition takes place, and the concepts of quantum field theory are essential tools in the detailed description of both processes. It is even more fruitful to regard electron–neutrino pair emission as an analogue to the emission (creation) of an electron–positron pair. An example of such a transition is $^{16}O^* \rightarrow {}^{16}O + e^+ + e^-$, where the initial and final states of the oxygen nucleus both have $J = 0$, and therefore cannot perform the transition by emitting a photon (recall p. 31). The transition can be depicted by the Feynman diagram

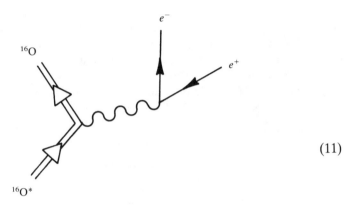

$$(11)$$

The wavy line in (11) represents the changing Coulomb field of the nucleus. This is a longitudinal field, in contrast to the transverse radiation field associated with photons; this Coulomb field can produce an e^+e^- pair in a state with $J = 0$. As we shall learn in §E.9, lepton pair creation via the weak interaction can be understood as being due to a mechanism that bears a profound resemblance to (11).

4. Lepton number conservation and parity violation

Of all the nuclear β-decays, the most fundamental is that of the isolated neutron

$$n \rightarrow p + e^- + \bar{\nu}, \tag{12}$$

which is Eq. (10^-) for $Q = 0$, $B = 1$. As we learned in §C.5, the process (12) is related by crossing to

$$p \rightarrow n + e^+ + \nu, \tag{13}$$

which is Eq. (10^+) for $Q = B = 1$. Since the neutron has a higher mass than the proton, only the first process (12) actually occurs for free nucleons. However, when nucleons are bound within nuclei, it often happens that replacement of a proton by a neutron decreases the total energy of the nucleus, in spite of the higher mass of the neutron. One reason for this is the decrease of electrostatic energy (the charge decreases by one unit); another is that in some cases the neutron is more strongly bound to the nucleus than the proton which it replaces.* Under these circumstances the reaction (13) occurs within the nucleus.

Another process related to the neutron's β-decay by crossing is

$$p + e^- \rightarrow n + \nu. \tag{14}$$

Here an electron bound in the atom combines with a proton bound in its nucleus to form a neutron and neutrino. As this electron capture is most probable when the electron occupies the shell closest to the nucleus—the so-called K-shell—(14) is called K-capture.

The final set of reactions related to (12) and (13) that we shall discuss provide the proof that $\nu \neq \bar{\nu}$. These are the neutrino scattering processes:

$$\nu + n \rightarrow p + e^-, \tag{15}$$

$$\bar{\nu} + p \rightarrow n + e^+. \tag{$\overline{15}$}$$

Nuclear reactors are copious sources of neutrinos because of the decay of the neutrons which abound in them. According to our definition, these are *antineutrinos* $\bar{\nu}$. If there is a distinction between $\bar{\nu}$ and ν, reactor neutrinos should *only* induce the reaction ($\overline{15}$), wherein a positron is emitted. If, on the other hand, there is no distinction between $\bar{\nu}$ and ν, reactor-produced neutrinos should cause both (15) and ($\overline{15}$), i.e., reactions in which electrons *or* positrons are emitted. Experiments have demonstrated that when targets are exposed to reactor-produced neutrinos, the only reactions observed involve the emission of e^+, thereby demonstrating that *the neutrino is distinct from the antineutrino.***

The cross sections for reactions induced by the low energy neutrinos produced by reactors are exceedingly small: of order 10^{-17} (fm)^2, corresponding to a mean-free-path of order 10^{14} km in solid rock! This should

---

* This statement seems to contradict our claim that the nuclear force does not distinguish between proton and neutron. The contradiction vanishes, however, when one considers the Pauli principle. It may be that the neutron after the process (13) finds a place in an orbital of lower energy which was occupied by protons, a state in which it could not have been when it was a proton before the decay.

** In detail, this argument runs as follows. Write (12) and (13) as $n \rightarrow pe\nu$ and $p \rightarrow n\bar{e}\nu'$, so that crossing implies $\nu p \rightarrow \bar{e}n$ and $\bar{\nu}'p \rightarrow \bar{e}n$. Since the latter process is observed, $\bar{\nu}' = \nu$, and therefore $\nu' = \bar{\nu}$.

bring home the meaning of the word "weak" in the name "weak interaction." As we shall see, neutrino cross sections grow with increasing energies; consequently, there is extensive data from accelerator-produced neutrino reactions in the subnuclear realm.

The distinction between ν and $\bar{\nu}$ allows us to formulate the law of *lepton number conservation*. If we assign the lepton number $N_l = 0$ to the nucleons, $N_l = 1$ to e^- and ν, and $N_l = -1$ to e^+ and $\bar{\nu}$, we see that all the processes discussed thus far conserve N_l. Apart from rather obvious modifications that are required by the existence of the other leptons mentioned previously, lepton number is, to our present knowledge, strictly conserved. Hence the lepton number is an "attribute" in the sense of our discussion on p. 43.

While the weak interaction has taught us this new conservation law, it has the astonishing property of violating reflection or mirror invariance. *As a consequence, parity is not conserved in weak interactions!* On p. 25 we stated the observational criterion for reflection invariance. We now apply it to the β-decay of the spin-polarized nucleus Co^{60}. Let $\hat{z}$ be the direction along which the spin is polarized, and consider a mirror in the $x–y$ plane perpendicular to $\hat{z}$ (see Fig. 4). If (p_x, p_y, p_z) is the momentum of a decay electron, its mirror image is $(p_x, p_y, -p_z)$. On the other hand, the mirror image of the initial state spin J_z is J_z, *not* $-J_z$, as one concludes by visualizing a rotation in the $x–y$ plane, or as one can see formally from the orbital angular momentum $xp_y - yp_x$. Consequently, reflection invariance requires equal probability P for electrons emitted at an angle of θ and $\pi - \theta$ with respect to the nuclear spin axis. The data show that $P(\theta) \neq P(\pi - \theta)$, or what is called an up-down asymmetry, and thereby demonstrate that spatial reflections are not a symmetry of the β-decay interaction.

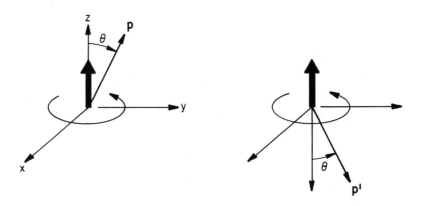

FIG. 4. Parity violation in β-decay. The two drawings shown differ by a reflection in the $x–y$ plane $\mathcal{P}$. The nuclear spin, indicated by the heavy arrow, is unaffected, as it generates a rotation in $\mathcal{P}$, but the electron momentum undergoes the change $\mathbf{p} \to \mathbf{p}'$. Hence, if reflection is a symmetry, the angular distribution of β-electrons must be up-down symmetric with respect to $\mathcal{P}$. It is not, and therefore parity is violated.

Summary

When we consider the world of atoms and nuclei, the "elementary" particles are protons, neutrons, electrons, neutrinos, and photons. We observe conservation laws, not only of energy, momentum, angular momentum, and charge, but also of other numbers: the number B of nucleons (protons plus neutrons), and the number N_l of leptons (electrons *plus* neutrinos). Both of these conservation laws hold in the sense that antiparticles must be counted as negative. Photons are not subject to a conservation law of numbers. They are bosons which are indistinguishable from their antiparticles, and can be created and absorbed singly, or in any number.

As for the forces that act between these "elementary" constituents of matter, we learn from nuclear physics that there are two interactions beyond the electromagnetic: the strong interaction between nucleons, responsible for binding nuclei, and the weak interaction, responsible for β-decay and other neutrino-related processes. While the strong interaction has all the symmetries familiar from atomic physics, the weak interaction is not reflection invariant.

It is remarkable that while subnuclear physics has exposed a deeper level of fundamental constituents, it has not revealed any new interactions between them.* On the other hand, by giving us a far better understanding of the constituents, it has permitted us to gain much more insight into the nature of the strong and weak interactions.

---

* See, however, *CP* violation [§E.11(b)], which may be due to a fundamentally new interaction. If proton decay is observed (§E.13), that would unambiguously demonstrate the existence of another new interaction.

E. SUBNUCLEAR PHENOMENA

1. Introduction

After the discovery that atomic nuclei consist of protons and neutrons, the composition of matter could be understood in terms of a very small number of elementary particles: the proton, the neutron, the electron, the neutrino, and the photon. However, this simple picture of the world did not survive for long; there were several early indications that the situation is more involved.

Studies of cosmic rays impinging upon the earth revealed the existence of new short-lived entities called baryons and mesons, which did not fit into this scheme. Furthermore, the observed interaction between protons and neutrons—the nuclear force—did not appear to be as simple and fundamental as the electromagnetic forces between charges, or magnetic moments. Being repulsive at short range and attractive at longer range, and dependent on the spin and the symmetry of the quantum state of the partners, it resembled the chemical force between atoms. As we have seen, the chemical force is really due to electrical forces between the constituents of atoms, and its complexity is a manifestation of the internal structure of atoms. The analogy with the chemical force suggests that the nuclear force is also a relatively complicated manifestation of more fundamental forces acting within the nucleon and connected with its internal structure.

The notion that the nucleon is an elementary particle received its final blow with the discovery of a short-lived excited state of the proton and the neutron, the so-called Δ. Another indication of internal structure came from the elastic scattering of electrons by nucleons. The angular distribution of the scattered electrons is directly related to the spatial charge distribution of the target (see §III.A.1, Vol. II). Measurements of this type revealed that nucleons have charge radii of ~ 0.8 fm, a dimension that is very similar to what one infers from pion-nucleon collisions, and from detailed models of hadronic spectra. Hence the nucleon cannot be "elementary," and has to have some internal structure.

This was the beginning of a series of discoveries that exposed a new realm of phenomena which comes into play if matter is subjected to energies of the order of several hundred MeV or more. It is a world full of short-lived entities such as these excited .nucleons, of mesons, and of heavier replicas of the electron. This plethora of particles (and antiparticles) are created and

59

transformed into each other in a variety of different collision and decay processes.

Despite this bewildering array of objects and phenomena, an underlying structure that is relatively simple has been discerned. The "particles" fall into three distinct families:

1. *The leptons*, of which the electron and its neutrino are the examples encountered in atomic and nuclear physics. To the best of our present knowledge, the "new" charged leptons are heavier replicas of the electron, and each has its own neutrino. All are spin-$\frac{1}{2}$ fermions.

2. *The quarks*, also spin-$\frac{1}{2}$ fermions. They combine together in various aggregates called *hadrons*, a name that is taken from the Greek word for "strong," because all these objects are subject to the strong interaction. The most familiar hadrons are the proton and neutron. These have a vast array of excited states, all having half-integer spin, of which the Δ just mentioned is the lowest-lying. In addition, there is a whole set of hadrons with integer spins; they are called *mesons*.

3. *Field quanta*, of which the photon—the quantum of the electromagnetic field—is the most familiar example. We now believe that the other basic interactions are also mediated by fields. The quanta of the strong field are called *gluons*; those of the weak field are called W^+, W^-, and Z^0; the first two carry electric charge, as the superscript indicates. They are massive particles whereas the photons and gluons are massless. All these field quanta are spin-one bosons—that is, *all the fields are vector fields*.*

Throughout most of this book we shall assume that hadrons are composed of quarks. But before introducing the quark model, we must present at least a sketch of the phenomena which this model seeks to encompass.

In trying to gain familiarity with the vast array of particles that are about to appear, one should not lose sight of one important and simplifying fact. As in nuclear physics, there are three basic interactions, the strong, the electromagnetic, and the weak. The strong interaction will turn out to be responsible for the internal dynamics of hadrons, and for the overwhelming part of the forces between hadrons. The characteristic time scale τ_h of hadronic dynamics can be estimated from the nucleon-Δ level spacing $\Delta E \simeq 300$ MeV: $\tau_h \sim (\Delta E)^{-1} \sim 10^{-23}$ sec. In comparison to this, all electromagnetic and weak phenomena proceed at a far slower pace, and can be ignored to an excellent approximation in any process caused by the strong interaction. But as we learned in nuclear physics, there are reactions and decays that can only be caused by the electromagnetic and weak interaction. They have small cross sections or long lifetimes, as the case may be. Those due to electromagnetism may violate isospin conservation, and those brought about by the weak interaction violate parity as well. These distinctions between the three interactions carry over into the subnuclear world.

---

* The quantum of the gravitational field, the graviton, is massless and is a spin-two boson. The field is a tensor field. It will not be considered in this book.

2. The baryon spectrum

We begin by describing the excited states which one observes when protons or neutrons are bombarded with particles having energies higher than several 100 MeV (see Fig. 5 and §III.A, Vol. II). Under such circumstances the nucleon can assume different properties that can be described as an excitation of the nucleon, analogous to the excitation of atoms or nuclei. Many excited states of the nucleon have been discovered. The characteristic energy differences are of the order of several hundred MeV.

Figure 6 shows the most important of these states as they are known today. We use the term "*baryon*" for the nucleon and its excited states, and reserve the term "nucleon" for the two lowest states, the proton and the neutron (p and n).

The baryon spectrum is the third level at which nature offers us a series of well-defined quantum states: the first were the atomic and molecular spectra, the second the nuclear spectra. The typical excitation energies are higher with each step: of the order of eV in the first, MeV in the second, and GeV in the third.

A study* of the states in the baryon spectrum shows that the same quantum numbers appear here as in atomic and nuclear spectra: the angular momentum J, and the isospin I. The appearance of multiplets characterized by J is no surprise: this tells us that the internal dynamics of the baryon is independent of the orientation of the coordinate frame in which the system is observed.

We assign a baryon number $B = 1$ to all these states, since we consider them to be excited states of the nucleon. B is identical to the quantity we used in §D to denote the number of nucleons in a nucleus.

The excited baryon states carry different electric charges Qe, all being integer multiples of the proton charge e. One of the most significant aspects of the baryon spectrum is the existence of almost degenerate levels of different charges which have all the characteristics of isospin multiplets. We already identified the nucleon as an isospin doublet $I = \frac{1}{2}$, with $I_3 = +\frac{1}{2}$ for the proton and $I_3 = -\frac{1}{2}$ for the neutron. Among the excited states, there are other nearly degenerate doublets with $Q = 1$ and $Q = 0$, which are also assigned to $I = \frac{1}{2}$. We also find different isospin multiplets. There are quartets of almost degenerate states, such as the Δ, to which we assign the isospin $I = \frac{3}{2}$, with $I_3 = -\frac{3}{2}$, $-\frac{1}{2}$, $+\frac{1}{2}$, $+\frac{3}{2}$ and $Q = -1$, 0, $+1$, $+2$, respectively, so that the charge formula, Eq. D(5), applies.

In addition to these doublets and quartets, there are also triplets and singlets, as well as doublets with charges that do *not* agree with Eq. D(5). A more general charge formula that encompasses all these nearly degenerate

---

* Some of the experimental techniques that yield the information used in this and the following section are summarized in §III.A (Vol. II); for a more detailed discussion, see Perkins (1982). Whenever we use data without a citation, our source is Particle Data Group (1982), or the compilation in Appendix I, where further references to data can be found.

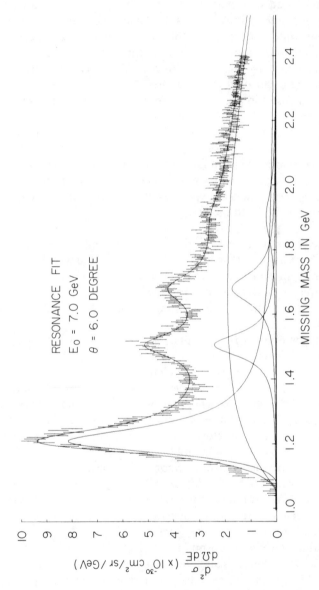

Fig. 5. Excitation spectrum of the proton as observed in the scattering of 7 GeV electrons through 6° (from SLAC-MIT Collaboration, 1968). The "missing mass" is the mass of the hadronic system, $\sqrt{(k + p - k')^2}$, where k and k' are the initial and final electron 4-momenta, and p the initial proton momentum (see §III.A.3, Vol II). The data have been fitted by four resonances, the most massive of which is barely discernable, and a smooth, nonresonant background. The three clearly visible resonances are $\Delta(1232)$, $N(1520)$, and $N(1680)$, in the notation defined in Table 1.

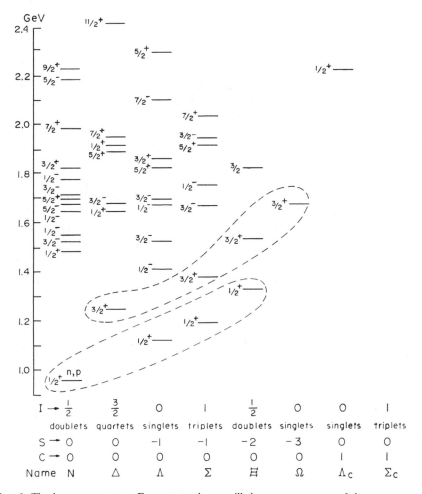

FIG. 6. The baryon spectrum. Every state shown will, by some sequence of decay processes, terminate in the proton. The numbers on the left of the levels indicate ordinary spin and parity. The eight columns of states differ in the quantum numbers I, S, and C as indicated. The last line gives the name of the group of states in each column. The lowest octet and decuplet [see §6(d)] are enclosed by broken lines.

multiplets is

$$Q = I_3 + \tfrac{1}{2}Y, \qquad Y = B + S + C, \tag{1}$$

where S and C are new quantum numbers ascribed to the baryon states. The quantum number S is called "*strangeness*," while the quantum number C is called "*charm*." The combination Y is referred to as *hypercharge*. Because of an unhappy convention, the strangeness quantum number of baryons is negative.

We already know that Q, B, and I_3 are additive quantum numbers—that if, say, $I_3^{(1)}$ and $I_3^{(2)}$ are the isospin projections for two objects, the combined system has the quantum number $I_3^{(1)} + I_3^{(2)}$. As they appear in the expression for Q, we assume that S and C are also additive.

The full significance of S and C will emerge presently. Their introduction is not just a trick for saving the charge formula. As we shall see, S and C are new attributes of elementary particles, exposed only at the subnuclear level, that have a certain kinship to the attribute of isospin revealed at the nuclear level. Like isospin, S and C will turn out to be conserved in all hadronic processes.[*]

Table 1 shows the states of the baryon grouped according to different values of isospin, S, and C. Table 1 lists the different groups, their quantum numbers, their isospin, the charges appearing in the multiplets according to Eq. (1), and the symbols by which the groups are named.[**] Figure 6 shows that each of these groups contains many baryons with different masses, and various spin-parities J^{Π}.

TABLE 1
Baryon states

Name	I	S	C	Multiplet	Charges
N	$\frac{1}{2}$	0	0	Doublets	$+1, 0$
Δ	$\frac{3}{2}$	0	0	Quartets	$+2, +1, 0, -1$
Λ	0	-1	0	Singlets	0
Σ	1	-1	0	Triplets	$+1, 0, -1$
Ξ	$\frac{1}{2}$	-2	0	Doublets	$0, -1$
Ω	0	-3	0	Singlets	-1
Λ_c	0	0	1	Singlets	1
Σ_c	1	0	1	Triplets	2, 1, 0

To each baryon level there should correspond an antibaryon level—the excited states of the antinucleon, $\bar{N}$. The antibaryon levels should have the same spectrum, the same spin J, and total isospin I, but the opposite charge, strangeness, charm, baryon number ($B = -1$), and projection I_3 of

---

[*] The discovery of a new type of meson, Y, demonstrates the existence of yet another quantum number beyond I, S, and C. We will discuss this evidence in §6(e). This quantum number will be called $\mathcal{B}$, and it is also borne by the so-called B-mesons of Fig. 8(c). Hence Eq. (1) should really read $Y = B + S + C + \mathcal{B}$. There are theoretical reasons to expect even a fifth quantum number at yet higher energy (see §12).

[**] As we see from Fig. 6, there are many levels with these quantum numbers. If one wants to be specific, one writes $\Delta(1232)$ or $\Sigma(1385)$, where the number in parentheses is the mass in MeV. When the symbols N, Δ, etc. of Table 1 are given without a mass, they usually refer to the lowest level. In this sense we have used Δ to refer to the lightest Δ-state, i.e., $\Delta(1232)$.

isospin. A large number of antibaryon levels having these expected properties have actually been observed.

We close this brief survey of the baryon spectrum with some qualitative remarks. In contrast to nuclei, and especially to atoms, the spacings in the baryon spectrum are comparable to the mass of the ground state (the nucleon). According to the argument on p. 31, this implies that any internal motions that give rise to baryon excitations are relativistic, i.e., have velocities comparable to c. This remark also applies to the mesons, which we shall discuss next. As a consequence, hadrons are relativistic systems, whereas atoms and nuclei are not.* For that reason, pair creation is a common occurrence in subnuclear physics, while it is a rare event in other branches of physics.

For a relativistic system, the level spacing is of order the internal momentum which, in turn, is of order R^{-1}. Since $\Delta E \sim 300$ MeV, the characteristic hadron radius R is $\Delta^{-1} \sim 10^{-13}$ cm, which is also of order $\tau_h c$. That hadrons actually have this size is confirmed by electron-proton and hadron-proton collision experiments (see §III.A, Vol. II).

3. Mesons

(a) Meson emission and absorption

We shall now discuss how the nucleon can be excited to higher states, and the way in which these return to lower states. Since all baryon levels have half-integer spin, any such transition requires an integer change of angular momentum. Consequently, the system that is emitted or absorbed by the baryon during such a transition can consist of any number of bosons, or of an even number of fermions, or of both.

As in atoms and nuclei, one observes the emission and absorption of photons in transitions between baryon states. These occur only between states of the same strangeness, and this tells us that there is a selection rule $\Delta S = 0$ for electromagnetic transitions.

Furthermore, as in nuclei, the weak emission of lepton pairs is observed in certain transitions from a higher to a lower state. In these transitions the strangeness may change by one unit, or remain unchanged, $\Delta S = 0$ or 1. In §9 we shall return to the interesting new features of weak processes in the subnuclear realm.

In nuclear spectra we encountered a new transition process, the emission of a lepton pair. Here, at the subnuclear level, yet another transition mode appears: the absorption or emission of particles belonging to a new species, the mesons. As mesons can be absorbed or emitted singly, it is clear that *mesons are bosons*.

---

* There are some important exceptions to this general rule, however; see §6(e) and §III.B (Vol. II).

If an excited baryon state can decay to a lower level by meson emission, the meson emission process will, in most cases, totally dominate the decay rate. Similarly, if a baryon level can be excited by meson absorption, that will have a far larger cross section than those for excitations by means of photoabsorption or by inelastic electron scattering,* while that for neutrino scattering will be vastly smaller still. All this merely illustrates our earlier statement that the mesons are hadrons, and participate in strong interaction processes, whereas photons and leptons do not.

(b) π-, K-, D-, and F-mesons

We begin our survey of mesons with those that are most long lived: the pions, kaons, D-, and F-mesons, denoted by π, K, D, and F, with appropriate superscripts that give their charge. A detailed compilation of mesons, and their properties, is provided in Appendix I.

Pions occur as a triplet, π^+, π^0, π^-, and are almost degenerate with a mass of about $\frac{1}{7}$ of the nucleon mass (see Table 2). The pions have zero spin and odd intrinsic parity, designated by 0^-. The π^0 decays into two photons with a lifetime of about 10^{-16} sec, whereas π^+ and π^- decay into a lepton pair with a lifetime of about 10^{-8} sec. The former is an electromagnetic, the latter a weak process, and both are enormously slow compared to the

TABLE 2
Pions, Kaons, D-, and F-mesons[a]

Symbol	Mass (MeV)	Q	I	I_3	S	C	Lifetime[b]
π^+	140	1	1	1	0	0	2.6(−8)
π^0	135	0	1	0	0	0	8.3(−17)
π^-	140	−1	1	−1	0	0	2.6(−8)
K^+	494	1	$\frac{1}{2}$	$\frac{1}{2}$	1	0	1.2(−8)
K^0	498	0	$\frac{1}{2}$	$-\frac{1}{2}$	1	0	c
$\bar{K}^0$	498	0	$\frac{1}{2}$	$\frac{1}{2}$	−1	0	c
K^-	494	−1	$\frac{1}{2}$	$-\frac{1}{2}$	−1	0	1.2(−8)
D^+	1869	1	$\frac{1}{2}$	$\frac{1}{2}$	0	1	~8(−13)
D^0	1865	0	$\frac{1}{2}$	$-\frac{1}{2}$	0	1	~4(−13)
$\bar{D}^0$	1865	0	$\frac{1}{2}$	$\frac{1}{2}$	0	−1	~4(−13)
D^-	1869	−1	$\frac{1}{2}$	$-\frac{1}{2}$	0	−1	~8(−13)
F^+	1970	1	0	0	1	1	~3(−13)
F^-	1970	−1	0	0	−1	−1	~3(−13)

[a] This table does not include the B-mesons; see Appendix I.
[b] Lifetimes are in seconds, and 2.6(−8) means 2.6×10^{-8} sec, etc.
[c] K^0 and $\bar{K}^0$ do not have well-defined lifetimes, but the linear combinations $(K^0 \pm \bar{K}^0)/\sqrt{2}$ do, and their lifetimes are 8.9(−11) and 5.2(−8). For a discussion of neutral kaons, see §11.

* On p. 64 we stated that I, S, and C are conserved in strong processes. Consequently mesonic processes only dominate if these conservation laws are satisfied.

characteristic hadronic time scale $\sim 10^{-23}$ sec. The K's, D's, and F's are also long-lived in this sense.

⟦The odd intrinsic parity of the pion is established by considering the reactions (a) $\pi^- d \to nn$, and (b) $\pi^- d \to nn \, \pi^0$; (a) *is* observed when low-energy π's strike deuterons; reaction (b) is *not* observed. The deuteron is made of a proton and a neutron in an S-state (orbital angular momentum $L = 0$), with total spin one. The pion is a spinless particle. Because of its low energy it will have no orbital angular momentum relative to the deuteron. Call Π_p, Π_n, Π_π the intrinsic parities of the proton, neutron, and pion. Then the parity of the initial state of both reactions is $\Pi_i = \Pi_p \Pi_n \Pi_{\pi^-}$. Because of angular momentum conservation, the two neutrons in the final state of the first reaction must have $J = 1$. The only antisymmetric nn state with $J = 1$ has $L = 1$, so that the parity of the final state of the first reaction $\Pi_f^{(a)} = -\Pi_n^2 = -1$. In the final state of (b), all particles have very low momenta since $m_{\pi^-} \simeq m_{\pi^0}$, and the deuteron binding energy is only 2 MeV. The overall $J = 1$ is therefore achieved by having as few particles in states with orbital angular momentum as possible: the nn pair can be in the $L = 0$ state, in which case π^0 has $l = 1$, or the nn pair can have $L = J = 1$, and π^0 has $l = 0$. In either case $\Pi_f^{(b)} = -\Pi_n^2 \, \Pi_{\pi^0} = -\Pi_{\pi^0}$.

Since the first reaction is observed but not the second, $\Pi_p \Pi_n \Pi_{\pi^-} = -1$, whereas $\Pi_p \Pi_n \Pi_{\pi^-} \ne -\Pi_{\pi^0}$. This requires π^0 to have odd intrinsic parity, whatever the parities of p, n, or π^-, a fact that has been confirmed by a study of polarization of the photons in the decay $\pi^0 \to 2\gamma$. We are then left with $\Pi_p \Pi_n = -\Pi_{\pi^-}$; we can therefore choose p and n to have the same parity (either even or odd), and π^- to have odd intrinsic parity like π^0. This is the usual, and most convenient choice. On the other hand, one could also use the awkward convention $\Pi_p = -\Pi_n$, $\Pi_{\pi^\pm} = 1$, which, if used consistently, does not lead to any contradictions. (In terms of quarks, this convention would assign opposite intrinsic parity to u and d.)⟧

It is natural to ascribe to the nearly degenerate π triplet a total isospin $I = 1$, so that π^+, π^0, π^- are assigned to $I_3 = 1, 0, -1$, respectively. If the charge formula (1) is to hold, Y must vanish for pions. We already know that all mesons have $B = 0$, or they could not be emitted in transitions between baryon levels. All pion-related reactions are also consistent with the assignment of $S = C = 0$ to π's, so that the charge formula does indeed hold. Since the pions' only attribute is isospin, charge conjugation has the effect of replacing I_3 by $-I_3$; hence π^+ and π^- are each others antiparticles, whereas π^0 (like the photon) is its own antiparticle.

A simple example of pion emission is the transition from the lowest Δ-state (mass 1232 MeV) to the nucleon N. As we saw, the Δ is an $I = \frac{3}{2}$ quartet with $Q = 2, 1, 0, -1$. Therefore, the transitions $\Delta \to N$ will be accompanied by the emission of pions of different charges. Figure 7 illustrates these transitions. This lowest-lying Δ has spin $J = \frac{3}{2}$, whereas N has $J = \frac{1}{2}$ and $I = \frac{1}{2}$. Indeed the pions carry away a unit of isospin, and also a unit of angular momentum. The latter comes from the fact that they are emitted in p-states (orbital angular momentum $l = 1$). The mean life of this lowest Δ is

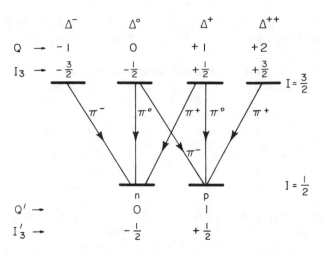

FIG. 7. Transitions from the four Δ-states to the two nucleon states by the emission of a pion.

0.4×10^{-23} sec. This corresponds to a width of 155 MeV, which is about half the Δ-N level spacing.

We now come to the *kaons*; there are four distinct K-mesons which form a nearly degenerate multiplet of about one-half the nucleon mass (see Table 2). As with the π's, they have zero spin and odd intrinsic parity. The charged K's decay into pion and lepton pairs, the neutral K's into pion pairs, with lifetimes in the 10^{-8} to 10^{-10} sec range. Although it is not immediately obvious, the pion pair decay modes are also weak processes (see §9).

The four K's have the charges 1, 0, 0, -1, and therefore it is not so clear how one is to assign their quantum numbers. A clue is provided by the observation that when a baryon emits or absorbs one K, its isospin always changes by half a unit, $\Delta I = \frac{1}{2}$. Consequently, we assign $I = \frac{1}{2}$ to the four K's, and that immediately implies that they form two isodoublets: $K \equiv (K^+, K^0)$, and their antiparticles $\bar{K} \equiv (\bar{K}^0, K^-)$, where in each doublet the first member has $I_3 = \frac{1}{2}$, the second $I_3 = -\frac{1}{2}$, in accordance with the definition of charge conjugation in §C.5.

If the charge formula (1) is to apply to the kaons, we must ascribe a quantum number beyond isospin to them. This role is played by strangeness; we assign $S = 1$ to the K-doublet, and $S = -1$ to the $\bar{K}$-doublet. The two neutral kaons, K^0 and $\bar{K}^0$, are therefore distinct, having $S = 1$ and -1, respectively. As mentioned before, strangeness has a significance that goes beyond the applicability of (1): as we shall learn in §4, it is a quantum number conserved in strong interaction processes.

The *D-mesons* bear a considerable resemblance to the K's. They also have spin zero and odd intrinsic parity, and the same charges, 1, 0, 0, -1. While they have the large mass $\sim$1870 MeV, they are also nearly degenerate, and have long lifetimes of about 10^{-13} sec (see Table 2). The distinction between D's and K's is that the former carry a unit of charm, and no strangeness, while the K's carry strangeness but not charm. The D's also are isotopic doublets ($I = \frac{1}{2}$). In addition, there are the F-mesons, with both strangeness and charm, but no isospin. Though they are isosinglets, they are charged ($F^{\pm}$).

Finally there are the very heavy B-mesons, which carry one unit of the quantum number $\mathscr{B}$, and like K and D are isodoublets.

(c) The meson spectrum

The nucleon is merely the ground state of a complex spectrum that we called the baryon spectrum. In an analogous manner, the π's are the lowest-lying members of the meson spectrum. As we saw, they have $J = 0^{-}$, $I = 1$, $S = C = 0$. Most other mesons with $S = C = 0$ are excited states of the pion; examples are $\rho(J = 1^{-}, I = 1)$, $\omega(J = 1^{-}, I = 0)$, and the other levels shown in Fig. 8. The K's also have excited states with the same strangeness ($S = 1$), the lowest being K^{*}, with $J = 1^{-}$ and $I = \frac{1}{2}$, and there is an analogous $J = 1^{-}$ excited state of the charmed meson D, called D^{*}.

All of these excited states can return to lower states by emitting pions:*
$\rho \rightarrow \pi\pi$, $\omega \rightarrow 3\pi$, $K^{*} \rightarrow K\pi$ and $D^{*} \rightarrow D\pi$. They do so with rates that are large, i.e., with lifetimes that are of the order of the hadronic time scale of 10^{-23} sec, and these are therefore transitions caused by the strong interaction.

In addition to these fast decays, there are very slow decays, such as $K^{+} \rightarrow \pi^{0}e^{+}\nu$ or $D^{+} \rightarrow K^{0}e^{+}\nu$, which are forbidden for the strong interaction, because they involve a change of S and C, respectively; indeed, these decays are due to the weak interaction. There are also rather faster electromagnetic decays, such as $\omega \rightarrow \pi^{0}\gamma$.

As with the spectra of the nucleus and the baryon, all these mesons are interconnected by some combination of strong, electromagnetic, or weak decays, and can be viewed as the excited levels of a single system, the meson.

4. Conservation of isospin, strangeness, charm, and baryon number

Nuclear physics has taught us that the isospin is not just a convenient way of labelling nearly degenerate groups of states; the total isospin I of a nuclear system is nearly a constant of motion because the nuclear forces are invariant under rotation in the isospin space (recall p. 50). The electromagnetic and weak interactions do not share this symmetry, however, and a nucleus can change its isospin by emitting a photon or lepton pair. All this carries over to the subnuclear realm, though with greater precision, since here we do not deal with systems like complex nuclei where the electrical forces are comparable to nuclear forces. Hence the conservation of isospin provides severe constraints on all reactions involving hadrons.

---

* We use the conventional abbreviation for reactions by writing $K^{*} \rightarrow K\pi$ instead of $K^{*} \rightarrow K + \pi$, $\pi N \rightarrow \Delta\rho$ instead of $\pi + N \rightarrow \Delta + \rho$, etc. Charges are often not shown explicitly, but charge conservation is always understood.

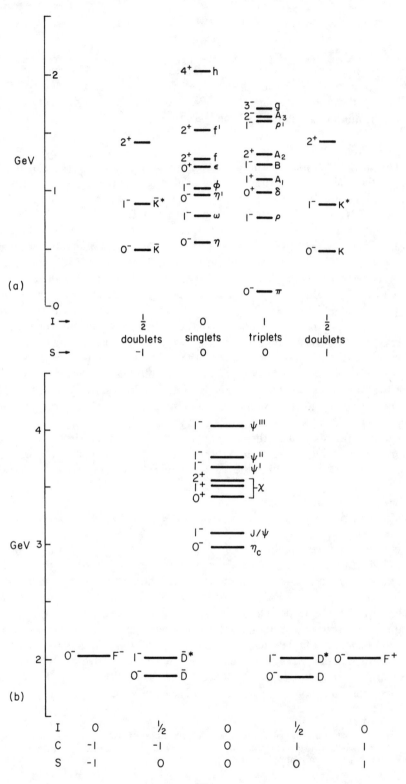

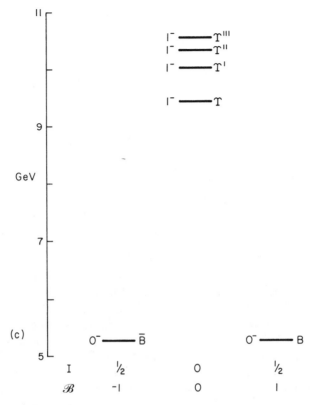

FIG. 8. The meson spectrum, with the quantum numbers displayed as in Fig. 6. Because of the large mass range, the figure is broken into three parts. The so-called "light" mesons are shown in (a). These have only the flavor I_3 and/or S, or neither; as we shall learn they are distinguished from the "heavy" mesons in (b) and (c) in that they only have the "light" quarks u, d, and s as constituents. Charmed mesons ($C = 1$ or -1), and the related $C = 0$ ψ family, are shown in (b); the former contain a c or $\bar{c}$ quark, the latter a $c\bar{c}$ pair. The heaviest known mesons are shown in (c). They too come in two varieties: The B-mesons, which have one unit of $\mathscr{B}$, and contain a b or $\bar{b}$ quark, and the Y-family, with $\mathscr{B} = 0$, composed of a $b\bar{b}$ pair.

Isospin conservation has many consequences for reactions and decays that are mediated by strong interactions*. One example is the absence of transitions from one Λ-baryon to another via pion emission, since the Λ's carry zero isospin. Other examples are quantitative relations (see §III.A.6, Vol. II) between the probabilities of pion emission in the various $\Delta(1232) \rightarrow N$ transitions shown in Fig. 7. These, and many other consequences of isospin conservation, have been verified experimentally.

---

* As we will see, the weak interactions can cause pion emission without conservation of I-spin or strangeness. However, these emissions occur with very small probability.

We shall now establish our claim on p. 64 that S and C are additive quantum numbers that are conserved in strong interactions. Since K's have $S = 1$, this would require that an initial $S = 0$ state, as in πN or NN collisions, could not lead to the production of a single K without other particles having $S = -1$. Thus $\pi N \to KN$ is forbidden, whereas $\pi N \to K\bar{K}N$ is allowed. The data confirm this: the former reaction is *never* observed, even in large statistics experiments, whereas the latter is no rarity. If we examine the baryon table (Table 1) we see that Λ and Σ are baryons with $S = -1$, and therefore $\pi N \to K\Lambda$ and $\pi N \to K\Sigma$ are allowed, and they indeed have appreciable cross sections.

One may ask whether conservation of S is really required to understand these facts: Since K and π have $I = \frac{1}{2}$ and 1, respectively, I-spin conservation forbids $\pi N \to KN$. The answer to this is provided by the two reactions

$$\pi^- p \to K^+ \Sigma^-, \tag{2}$$

$$\pi^- p \to K^- \Sigma^+. \tag{3}$$

Here the states have half-integer isospin, so isospin conservation cannot rule out either process. On the other hand, $\Sigma^\pm$ both have $S = -1$, whereas $K^\pm$ have $S = \pm 1$, so that (2) conserves S, whereas (3) does not. Once again the data show that (2) is a copious process, whereas (3) never occurs. This is called the phenomenon of *associated production*: if the initial state (e.g., πN or NN) has zero strangeness, any produced strange particle must be accompanied by another of opposite strangeness. Figure 9 shows an example of associated production, as well as the subsequent weak decays of the strange particles produced in the collision.

Conservation of strangeness forbids any decay by pion emission that leads to a change of strangeness, such as $\Lambda \to \pi N$, $\Xi \to \pi \Lambda$, and $\Omega \to \pi \Xi$. These examples are also forbidden by isospin conservation since I changes by half a unit. However, transitions such as $\Xi \to N\pi$ would not be forbidden by I-conservation, but are excluded because $\Delta S = 2$. But conservation of S allows decays such as $\Lambda \to \bar{K}N$, $\Sigma \to \bar{K}N$, $\Xi \to \bar{K}\Lambda$, and $\Omega \to \bar{K}\Xi$. Indeed, these $\bar{K}$-emissions occur copiously for excited Λ, Σ, and Ξ's (no excited Ω has yet been found), but not for the lowest ones. Among the lowest baryon states N, Λ, Σ, Ξ, and Ω, the pairs of states that differ by one unit of S have a difference in mass smaller than the K-mass. Therefore these lowest baryon levels are stable with respect to the strong interaction, and only decay slowly via the weak interaction.

Note the difference between forbidden processes that occur as collisions, like $\pi N \to KN$, and those that occur as decays, like $\Lambda \to \pi N$. Collision processes mediated by weak interactions, such as $\pi N \to \Lambda \pi$, are usually unobservable because of the enormous background from allowed hadronic processes. On the other hand, the weak decay of a particle that cannot decay via the strong interaction is easily observable by waiting long enough.

FIG. 9. A bubble chamber photograph of associated production, $\pi^- p \to K^0 \Lambda^0$, followed by the decays $\Lambda^0 \to p\pi^-$ and $K^0 \to \pi^+ \pi^-$. The incident π^- track is indicated by the arrow. This event took place in the Berkeley 10 in. hydrogen chamber (Crawford, 1957).

Hadrons that only decay weakly can be seen in a bubble chamber (see Figs. 9 and 33). When they are charged they leave a visible track before they decay; when they carry no charge, the decay takes place far from the point of creation. One can even make beams of weakly decaying particles and study their collisions before they decay (e.g., π, K, Λ). Hadrons that decay strongly don't live long enough to leave a track, or to form a beam.

We shall not say much about the evidence for conservation of charm in strong interactions. The D-meson may suffice as an example. The most striking property of D is that it is long lived ($\sim 10^{-13}$ sec) despite the large number of available decay channels made accessible by its large mass. For example, D cannot be an $S = 1$ isodoublet, for then it would decay strongly into $K\pi$. It therefore is necessary to assign to D a quantum number that is *not* carried by pions or kaons, and it must be conserved in strong interactions. This quantum number is charm.

To summarize, I, S, and C are conserved in strong but not in weak processes. The observed decays $\Lambda \rightarrow N\pi$, $K \rightarrow \pi\pi$, and $K \rightarrow$ lepton pair, all violate conservation of S and I, while $D \rightarrow K\pi$ and $D \rightarrow K +$ lepton pair violate C. All these decays have lifetimes enormously long compared to 10^{-23} sec.

Finally, we come to conservation of baryon number B. Insofar as we now know, this is absolutely conserved, for if it were not, the proton could decay as, for example, in $p \rightarrow e^+ \pi^0$, or $p \rightarrow e^+ \nu\nu$. The current lower limit on the proton lifetime is $\sim 10^{31}$ years, which is far longer than the lifetime of the universe. Nevertheless, there is now an extensive search for proton decay stimulated by some intriguing theoretical conjectures (see §13).

5. Quarks

(a) General remarks

We believe today that the great complexity of hadronic phenomena is due to the fact that hadrons, like atoms or nuclei, are composite structures built up from a small number of structureless objects called *quarks*. At first it was thought that these new "elementary" particles only provided an elegant and compact way of classifying hadronic states, but mounting evidence indicates that the quark model has a deeper dynamical significance. Virtually all of hadronic physics—spectroscopy, weak and electromagnetic decays, collisions with electrons or neutrinos, and hadron production in e^+e^- annihilation— can be understood to a remarkable degree in terms of the quark model.

Strong evidence for the existence of quarks inside the nucleon is provided by a group of experiments called deep inelastic scattering.* In these experiments, electrons, muons, and neutrinos of high energy (15 to 200 GeV) collide with nucleons, and one selects "deep inelastic" events, i.e., where a large amount of momentum and energy is transferred to the nucleon. The incident particles are scattered through angles much larger than expected from a continuous distribution of charge within the nucleon. This result can be interpreted as due to the presence of objects within the nucleon, having a

* For a detailed treatment see Chap. V, Vol. II.

size far smaller than it, just as the presence of an atomic nucleus was inferred from the large-angle scattering of α-particles. The wavelength of the incident beam sets the resolving power; at this time the upper limit on the size of these constituents of the nucleon is of order 10^{-16} cm, or $\sim 10^3$ times smaller than the nucleon's dimensions. These experiments also show that these constituents have spin $\frac{1}{2}$. Thus charged point-like fermions have been "seen" within the nucleon, in the sense that any seeing consists of an analysis of particle collisions, such as of photons in the case of visual observations.

Another development that has lent great credence to the quark model was the discovery of two families of mesons that have an excitation spectrum that is quite similar to that of the hydrogen atom. These are the ψ and Υ families, shown in Fig. 8. As we shall see in §6(e) and §III.B (Vol. II), the quark model can account for these spectra in quantitative detail.

The theory of the strong, electromagnetic, and weak interactions also becomes more elegant and powerful when it ascribes the basic hadronic degrees of freedom to quarks.

The remainder of this book treats hadrons in terms of the quark model. It is by no means certain that the quark model, in its present form, represents the ultimate theory of hadronic structure. Many questions remain open, the most pressing probably being whether quarks exist in isolation, or only bound inside hadrons. But the model is so successful that many of its concepts are expected to survive in the ultimate theory.

(b) Basic assumptions

1. Quarks, denoted by q, are spin-$\frac{1}{2}$ particles. It is necessary that at least some q's be fermions, since we must be able to combine them into objects of half-integer spin (baryons), and also of integer spin (mesons). The assumption that all q's have $J = \frac{1}{2}$ is the most economical.

2. The baryon is composed of three quarks, written as qqq. The baryon spectrum corresponds to the various quantum states of the qqq system. Since the baryon number B is additive, we assign $B = \frac{1}{3}$ to all quarks. Antiquarks $\bar{q}$, with $B = -\frac{1}{3}$, are therefore distinct from quarks.

3. The meson is composed of a quark and antiquark, $q\bar{q}$, and therefore has $B = 0$. The meson spectrum corresponds to the states of the $q\bar{q}$ system. As its constituents can annihilate, this accounts for the fact that there are no absolutely stable meson states.

4. Hadrons are characterized by the additive quantum numbers I_3, S, and C. As hadrons are either qqq or $q\bar{q}$ composites, we must assign appropriate values of I_3, S, and C to quarks. The most economical assignment is to assume five distinct quark types:

   (a) A pair of quarks, u and d, that carry only isospin, but none of the other quantum numbers. Thus (u,d) are an $I = \frac{1}{2}$ doublet, with u standing for "up," or $I_3 = \frac{1}{2}$, and d for "down," or $I_3 = -\frac{1}{2}$; $(\bar{d}, \bar{u})$ is the antidoublet. All have $S = C = 0$. One often refers to u and d as "ordinary" quarks.

(b) The strange quark s, which carries strangeness, but not charm or isospin. It is most convenient to assign $S = -1$ to s.

(c) The charmed quark c, which carries charm, but not strangeness or isospin, and to which one assigns* $C = 1$.

(d) A fifth** quark b, which carries none of the quantum numbers listed thus far (i.e., $I = S = C = 0$), but has yet another attribute, $\mathcal{B}$. This quark is required by the Y-family, and the B-mesons, shown in Fig. 8(c). It is very important in a restricted, though significant, set of phenomena to be discussed in §6(e) and §III.B (Vol. II), but it is irrelevant to most of hadronic physics.

Table 3 compiles these assignments, as well as other properties of quarks that will be discussed below. It has become customary to use the term *flavor* for the attributes I_3, S, and C that distinguish the various quarks.[†] The antiquarks have the opposite flavor. Thus $q\bar{q}$ states such as $u\bar{u}$ or $c\bar{c}$ (and linear combinations of any of these) are *flavor-neutral* states. At times we shall also use the notation q_i, where the subscript denotes one of the flavors u, d, s, or b; $\bar{q}_i$ then has the opposite flavor.

TABLE 3
Quarks

Type (flavor)	up	down	strange	charmed
Symbol	u	d	s	c
Spin	$\frac{1}{2}$	$\frac{1}{2}$	$\frac{1}{2}$	$\frac{1}{2}$
Charge	$\frac{2}{3}$	$-\frac{1}{3}$	$-\frac{1}{3}$	$\frac{2}{3}$
Isospin I	$\frac{1}{2}$	$\frac{1}{2}$	0	0
I_3	$\frac{1}{2}$	$-\frac{1}{2}$	0	0
Strangeness	0	0	-1	0
Charm	0	0	0	1
Baryon number	$\frac{1}{3}$	$\frac{1}{3}$	$\frac{1}{3}$	$\frac{1}{3}$
Effective mass (approx.) in MeV	m_0	m_0	$m_0 + 150$	$m_0 + 1500$

Here we do not list the fifth quark b, which is firmly established. It has an effective mass ~5 GeV, and $Q = -\frac{1}{3}$, and is contained in the mesons shown in Fig. 8(c). A sixth quark, t, is expected on theoretical grounds (see §12); at this time (1983) we know that $m_t \gtrsim 20$ GeV. The masses m_0 of u and d are probably very small compared to the proton mass.

---

* It is a historical accident that s has negative strangeness, whereas c has positive charm.
** There are strong theoretical arguments for the existence of a sixth quark, called t; see §12.
† In this language, the established b- and conjectured t-quarks carry two new flavors; that carried by b is called $\mathcal{B}$ by us, and has the value 1 for b.

(c) Quark charges and masses

The additive nature of all quantum numbers in the charge formula (1) means that it must apply to the quarks themselves.* Substituting $B = \frac{1}{3}$, and our assignments (a)–(c) in the preceding list, into the charge formula (1) yields

$$Q_u = \tfrac{1}{2} + \tfrac{1}{2} \cdot \tfrac{1}{3} = \tfrac{2}{3},$$
$$Q_d = -\tfrac{1}{2} + \tfrac{1}{2} \cdot \tfrac{1}{3} = -\tfrac{1}{3}, \tag{4}$$
$$Q_c = 0 + \tfrac{1}{2}(\tfrac{1}{3} + 1) = \tfrac{2}{3},$$
$$Q_s = 0 + \tfrac{1}{2}(\tfrac{1}{3} - 1) = -\tfrac{1}{3}.$$

Thus quarks have fractional charges whereas, by construction, all baryons $q_i q_j q_k$, and all mesons $q_i \bar{q}_j$, have integer eigenvalues of Q. This remarkable property of quarks has stimulated many searches for fractionally charged particles. So far these searches have failed, with the exception of one experiment that has reported fractional charges on macroscopic bodies.**

Since only the ordinary quarks u and d carry isospin, all isospin properties of hadrons must come from these quarks. By the same token, any property of a hadron that depends on its strangeness or charm must be due to its having s or c quarks as constituents.

A few words are necessary concerning the masses of quarks. The mass of a particle can only be precisely defined by its energy when it is a free particle. Since isolated quarks have not been observed, it is difficult to give a precise meaning to their mass. As we will see, the mass *differences* between quarks are more meaningful, though even here some ambiguities will remain. That is why Table 3 indicates only approximate mass differences.

The problem of the quark mass is somewhat analogous to phenomena that arise in condensed matter physics. An electron moving through a solid behaves as if it had an "effective" mass m_{eff} that can differ appreciably from the true mass m. Furthermore, m_{eff} may depend on the particulars of the motion, because $(m_{eff} - m)$ is really due to interaction between the electron and the bodies in its neighborhood. By the same token, all masses of quarks are "effective," since they are ascertained from phenomena where the quark is interacting with other quarks. For that reason, the mass that emerges from an analysis of the hadron level scheme may differ appreciably from that which is appropriate to, say, a weak decay. Despite these caveats, several things are clear: the mass of the c-quark, m_c, is far larger than that of u, d, or s, and the s quark is rather heavier than u or d. There is considerable ambiguity attached to m_u and m_d; these quantities are sensitive to details of the model

---

* Note that the conclusions presented here depend on our "minimal" assumptions. More complex schemes, involving a larger number of quarks, can be constructed so that all Q's are integers.

** LaRue et al. (1981).

that are not fully established, though there is reason to believe that these masses (called m_0 in Table 3) are very small compared to the proton mass.

(d) Interactions

As soon as one views hadrons as quark aggregates, it becomes necessary to specify how quarks participate in the strong, electromagnetic, and weak interactions. We have already learned (§4) that all flavor quantum numbers are conserved in the strong interaction. This global property of hadrons is guaranteed if we make the following assumption concerning the behavior of quarks: *In a strong interaction process, a quark cannot change its flavor; furthermore, when the strong interaction causes the creation or destruction of quark–antiquark pairs, these must be flavor neutral.*

It is an established experimental fact that *the electromagnetic interaction also conserves all flavors.** For example, the strangeness conserving decays $\Sigma^0 \rightarrow \Lambda^0 \gamma$ and $K^* \rightarrow K\gamma$ are observed, but not the $\Delta S = 1$ transitions $\Sigma \rightarrow N\gamma$ and $K \rightarrow \pi\gamma$. The same applies to charm: $D^* \rightarrow D\gamma$ is observed but not the $\Delta C = 1$ transitions $D \rightarrow \pi\gamma$ and $D \rightarrow K\gamma$. This means that the quark transitions $c \rightarrow u\gamma$ and $s \rightarrow d\gamma$ are strictly forbidden, even though they violate no principle of electrodynamics. The preceding statement concerning the lack of change in the flavor distribution in a strong interaction process therefore holds verbatim for the electromagnetic interaction as well. The weak interaction is different, however; as we saw in §4, the flavor carried by hadrons sometimes (but not always) changes in a weak process. At the quark level this means that a quark may change its flavor in reactions caused by the weak interaction.

There must exist strong forces between quarks which bind them together into hadrons, and which result in the strong interactions between hadrons. For example, the nuclear force must be a consequence of the forces between the quarks inside neutrons and protons, just as the chemical force is a consequence of the electrical forces between electrons and nuclei. We shall have much more to say about these forces in §7 and Vol. II, Chap. IV. For the time being it is only necessary to make the statement that *the strong forces between quarks are independent of the quark flavors.*** In particular, the members of the isodoublet (u,d) therefore have the same strong interactions. The u–d mass difference is small compared to all hadron masses, so there is virtually complete symmetry between the members of this doublet. *The observed isospin invariance of the strong interaction is an important consequence of this symmetry.*†

There is another symmetry, though not exact, that is also of considerable importance in hadronic spectroscopy. The mass splitting between s on the one

---

* The electromagnetic interaction conserves I_3, but not I, as discussed in §8(a). This is in keeping with our definition of flavor in §5(b).

** Note that while the strong and electromagnetic interactions both conserve flavor, the latter is flavor dependent, since the flavor of a quark determines its charge.

† Flavor independence is a consequence of the invariance principle that underlies QCD [see §7(a)], but the small u–d mass difference is not understood at this time.

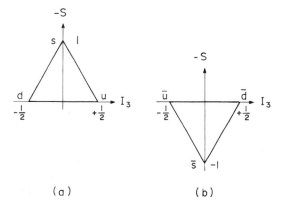

FIG. 10. (I_3, S) diagrams for the u, d, s quarks; (a) for quarks, (b) for antiquarks.

hand, and (u,d) on the other, is only of order 150 MeV, and this is small compared to typical hadron masses. Hence the triplet (u,d,s) can, to a reasonable first approximation, be viewed as degenerate in many circumstances. This leads to the so-called flavor $SU(3)$ symmetry, which we will exploit in §§6 and 7. On the other hand, the c- and b-quarks are so massive that one can only treat them on an equal footing with the lighter three in very high-energy collisions. In spectroscopy, c and b stand apart from the others, and systems composed of such heavy quarks (especially $c\bar{c}$ and $b\bar{b}$) must be treated separately, which we shall do in §6(e). For that reason, the existence of c and b will usually be ignored in the following.

It is useful to represent the three light quarks u, d, s (and all hadron states composed of them) in a graphical scheme. A two-dimensional diagram suffices, since each of the three quarks is specified by two "coordinates": I_3 and S (or Y). Such a diagram is shown in Fig. 10(a), where $-S$ is plotted upward for convenience, and the isodoublet (u,d) lies on a horizontal line. Note that the choice of scales is such that an equilateral triangle results, i.e., a perfect threefold symmetry. Such diagrams are particularly useful in the approximation where the three quarks are assumed to have equal masses.

Figure 10(b) shows the triplet of antiquarks, i.e., the charge conjugates of the quarks in Fig. 10(a). The threefold symmetry is maintained. The transformation of $q_i \to \bar{q}_i$ is seen to be represented by an inversion through the center ($I_3 = 0$, $S = 0$) of the diagram, in accordance with our definition of charge conjugation in §C.5.

6. Hadronic spectroscopy

(a) Baryon and meson types

We now draw some simple conclusions from the hypothesis that baryons consist of three quarks. If the three quarks are all of the u- and d-type, the resulting baryon will carry the quantum numbers $S = C = 0$. If one, two, or three of the quarks are of the s (or c) type, the baryon will have $S = -1, -2, -3$ (or $C = 1, 2, 3$), respectively. No baryon with strangeness less than -3, or charm more than 3, is possible. The larger $|S|$ or C, the larger should be the mass, since s and c are supposed to be heavier than u or d. (This should be more noticeable in the case of charm.) All these conclusions are indeed fulfilled, although very little is known about charmed baryons, except that they are very massive.

The hypothesis that three quarks constitute a baryon also explains the existence of the six groups of baryon levels with $C = 0$ in Table 1. When $S = C = 0$, all three quarks must be of the u- and d-type, having $I = \frac{1}{2}$; the combination of three isospins $\frac{1}{2}$ gives either $I = \frac{3}{2}$ (Δ-baryons) or $I = \frac{1}{2}$ (N-baryons). When $S = -1$, two quarks are of the ($I = \frac{1}{2}$)-variety, giving $I = 1$ (Σ-baryons) and $I = 0$ (Λ-baryons). When $S = -2$, only one quark has $I = \frac{1}{2}$, so that the Ξ-baryons have $I = \frac{1}{2}$. When $S = -3$, no constituents carry isospin, giving the Ω-baryons with $I = 0$.

It is equally easy to predict the baryon groups for $C \neq 0$. Only baryons with $C = 1$ have been found thus far. In the $S = 0$ case two quarks are of the ($I = \frac{1}{2}$)-variety; they combine with the third c-quark to give Σ_c-baryons with $I = 1$, $C = 1$, and Λ_c-baryons with $I = 0$, $C = 1$, in analogy to the Σ- and Λ-baryons, where the third quark is s.

All baryon groups in Table 1 are now accounted for. Baryons with $C = 2$ or 3, presumably exist but have not yet been observed. There is a systematic increase of mass as (u,d)-quarks are replaced by s-quarks, and a large jump if the baryon contains a c-quark (see Fig. 6).

Similar conclusions can be drawn from the hypothesis that mesons are quark–antiquark pairs. The quantum numbers of mesons with $C = 0$ in Table 2 are readily understood. It follows immediately that the nonstrange mesons must have $I = 1$ or 0, since they consist of two $I = \frac{1}{2}$ quarks. We have a strange meson when one of the constituents is s or $\bar{s}$, in the first case a meson with $S = -1$, in the second with $S = +1$. Both belong to isodoublets, because of the nonstrange quark with $I = \frac{1}{2}$. When both constituents are strange, the meson has $S = 0$ and $I = 0$, since s and $\bar{s}$ have opposite strangeness. Thus we never get mesons with $|S| > 1$, as is borne out by the facts.

The same holds also for charm. There will be heavy mesons with $S = 0$, $C = 1$, namely $c\bar{u}$ and $c\bar{d}$, which form an isospin doublet. Their antiparticles, with composition $u\bar{c}$ and $d\bar{c}$, form an isodoublet with $S = 0, C = -1$. Mesons carrying both strangeness and charm have the composition $c\bar{s}$ and $\bar{c}s$, and therefore are isosinglets with $Q = 1$ and -1 respectively; at the moment the only known examples are F^+ and F^- (see Table 2).

There are also heavier mesons with $S = C = 0$, consisting of a charmed quark and a charmed antiquark. Examples are J/ψ, χ, η_c, etc., shown in Fig. 8. The Υ's have the composition $b\bar{b}$, where b is the most massive quark presently known (Fig. 8).

We now discuss hadron structure in greater detail. In the remainder of this section we will confine ourselves to those hadrons which consist of the three *lightest* quarks, i.e. the u, d, and s-types. These should represent the lowest lying hadrons, an expectation that is confirmed by the observed hadron spectra.

(b) Meson nonets

We begin with the mesons. Let the lowest quantum state be $\psi(q\bar{q})$. Various choices of flavor for q and $\bar{q}$ will produce different mesons. What kind of states do we get? Assuming the wave function $\psi(q\bar{q})$ to be symmetric in the space coordinates, as we would expect for the lowest states, the parity of all the states $\psi(q\bar{q})$ must be *odd*, since any fermion–antifermion system has odd intrinsic parity (recall §C.6). Hence we obtain *two* sets of states, those where the spins of the quarks are parallel $(J = 1^-)$, and those where they are antiparallel $(J = 0^-)$. The first are called *vector mesons*,* the second *pseudoscalar mesons*.

Mesons consisting of u and d (those with zero strangeness) correspond to the combinations given in Table 4, together with their charges and isospins. The charges are obtained by adding the fractional charges of the quark and antiquark. Since the u and d are the members of an isospin doublet, the quark pair will have $I = 1$ or $I = 0$. The three substates of $I = 1$ are $u\bar{d}$, $(u\bar{u} - d\bar{d})/\sqrt{2}$, $\bar{u}d$. The analogy with the p,n pair in Eq. D(8) is not complete since we now deal with a particle–antiparticle pair, whereas both p and n are particles. Since the antiparticles have I_3's opposite to the particles, it is the combinations $u\bar{d}$ and $d\bar{u}$ which have $I = 1$, $I_3 = \pm 1$, since each partner has the same I_3. Another change from Eq. D(8) is found in the $I = 1$, $I_3 = 0$ state.** In the particle-antiparticle case it is the combination

---

TABLE 4

Light mesons without s-quarks

Combination	$d\bar{u}$	$d\bar{u}$	$2^{-\frac{1}{2}}(u\bar{u} - d\bar{d})$	$2^{-\frac{1}{2}}(d\bar{d} + u\bar{u})$
Charge	$+1$	-1	0	0
Isospin I	1	1	1	0
I_3	$+1$	-1	0	0
Name: $J = 0$	π^+	π^-	π^0	η
$\quad\quad\quad J = 1$	ρ^+	ρ^-	ρ^0	ω

---

* An ordinary vector field (the electric field is an example) has quanta with spin 1 and odd parity.

** The argument that the $(I = 1, I_3 = 0)$ state must be symmetric [see Eq. D(8)] does *not* hold here since $u\bar{d}$ and $d\bar{u}$ are not symmetric either. Many authors use a phase convention in which the $I_3 = -1$ state is $-\bar{u}d$.

$(d\bar{d} + u\bar{u})/\sqrt{2}$ which is invariant under rotations, and therefore this is the $I = 0$ state; the orthogonal state $(u\bar{u} - d\bar{d})/\sqrt{2}$ has $I = 1$, $I_3 = 0$.

The four combinations in Table 4 represent the isotriplet $\pi^+\pi^0\pi^-$, and the isosinglet η, when the spins are opposed (pseudoscalar mesons). When the spins are parallel (vector mesons) they represent the isotriplet $\rho^+\rho^0\rho^-$, and the isosinglet ω. Since the quark interactions are invariant under isospin rotations, the components of the isotriplets should have essentially the same mass. This is indeed true for the three pions and the three ρ-mesons. The small mass differences, of the order of a few MeV, can be understood as being due to their differing electromagnetic energy.

Next consider mesons in which one of the quarks or antiquarks is strange. There are five such combinations, as listed in Table 5. The isospin comes solely from the u or d quark. Again the members of the isodoublets have almost equal mass (see Table 2). The charges of the mesons in Table 4 and Table 5 fulfil the relation in Eq. (1), since $B = 0$ for mesons.

TABLE 5
Light mesons containing s-quarks

Combination	$u\bar{s}$	$d\bar{s}$	$s\bar{u}$	$s\bar{d}$	$s\bar{s}$
Charge	$+1$	0	-1	0	0
Isospin	$\frac{1}{2}$	$\frac{1}{2}$	$\frac{1}{2}$	$\frac{1}{2}$	0
I_3	$+\frac{1}{2}$	$-\frac{1}{2}$	$-\frac{1}{2}$	$+\frac{1}{2}$	0
Strangeness	$+1$	$+1$	-1	-1	0
Name: $J = 0$	K^+	K^0	K^-	$\bar{K}^0$	η'
$J = 1$	K^{*+}	K^{*0}	K^{*-}	$\bar{K}^{*0}$	ϕ

Altogether we have obtained nine $q\bar{q}$ combinations, a *nonet*, with antiparallel spin $(J = 0^-$, pseudoscalar mesons), and another nonet with parallel spins $(J = 1^-$, vector mesons). This accords well with the observed low-mass meson spectrum. The pseudoscalar nonet comprises the pion isotriplet, the K and $\bar{K}$ isodoublets, and the two isosinglets η and η'. The vector nonet contains the ρ-triplet, the K^* and $\bar{K}^*$ doublets, and the singlets ω and ϕ.

In each nonet there are three linearly independent states having $I_3 = S = 0$: $u\bar{u}$, $d\bar{d}$, $s\bar{s}$. If the quark masses were equal, these would be degenerate, and therefore small perturbations can cause large mixings. There are two such perturbations: the quark mass differences themselves, and the spin-dependent forces that raise the $J = 1^-$ states above the 0^-. In the case of the vector nonet, it turns out that the mass difference is the decisive perturbation. It leads to eigenstates for vector mesons with $I_3 = S = 0$ as given in Tables 4 and 5, where $\phi = s\bar{s}$ is one of the eigenstates. In the

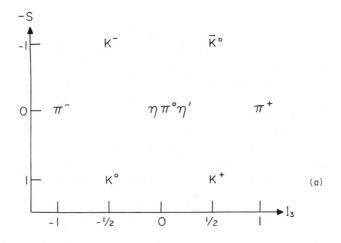

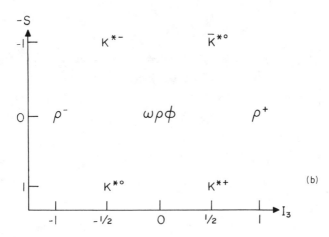

FIG. 11. (I_3, S) diagrams of mesons; (a) the nonet with zero spin (pseudoscalar mesons); (b) the nonet with spin 1 (vector mesons).

pseudoscalar nonet, the eigenstates for η and η' have a somewhat different linear combination.*

Figure 11 shows the 0^- and 1^- nonets on the I_3 vs. S plot used previously for quarks (Fig. 10). These diagrams have several interesting features. They are hexagons, i.e., have threefold symmetry, reflecting the equivalence of the three quarks of which the mesons are composed. Isomultiplets lie on

---

* There are $|\eta\rangle = 6^{-\frac{1}{2}}(u\bar{u} + d\bar{d} - 2s\bar{s})$, and $|\eta'\rangle = 3^{-\frac{1}{2}}(u\bar{u} + d\bar{d} + s\bar{s})$. The approximate symmetry between u, d, and s (SU_3 flavor symmetry) mentioned on p. 79 leads to these states. It was once customary to split the meson into an octet and a singlet. This is natural if flavor SU_3 symmetry is used to construct the basic states, but in most cases the quark basis works better, in which case the octet-singlet split is somewhat artificial.

horizontal lines, e.g., the K^0K^+ doublet and the $\pi^+\pi^0\pi^-$ triplet. Horizontal displacement by one unit corresponds to substituting a u quark by a d quark (or vice versa). Similarly, displacement parallel to the tilted sides corresponds to (u,s) and (d,s) interchange, as is already evident in the quark diagrams, Fig. 10. Mesons of the same charge Q (say K^+ and π^+) lie always on the same line at 60° to the I_3 axis. One point of the diagrams, the origin at $I_3 = S = 0$, has "triple occupancy": π^0, η, η', and ρ^0, ω, and ϕ. This is the previously mentioned threefold degeneracy for $I_3 = S = 0$. The antiparticles of the mesons appear always at diametrically opposite points of the figure. The three mesons at the center are their own antiparticles.

The masses of these mesons are plotted in Fig. 12. The mesons containing ordinary quarks are always lightest. Substitution of s quarks raises the mass.

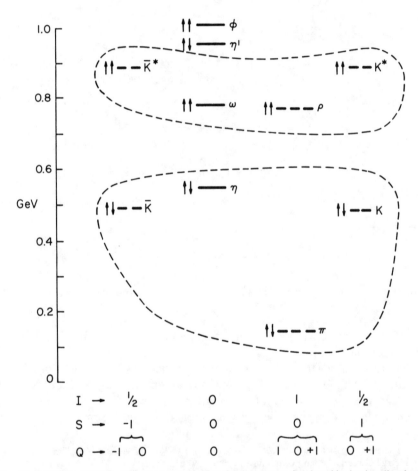

Fig. 12. Spectrum of the lowest meson states, showing two octets (enclosed by broken lines) and two singlets. Together they form the pseudoscalar (antiparallel quark spins ↑↓) and vector (↑↑) meson nonets.

Obviously strong spin–spin forces must exist, since ρ mesons are so much heavier than π mesons, although they differ only in the relative direction of the quark spins. The mass differences between mesons containing s-quarks, and those that do not, are not necessarily equal to $m_s - m_0$; the higher mass of the s-quark makes its motion less relativistic, and that also has an influence on the energy of the meson.

This completes our discussion of the lightest meson states. All observed states have precisely the quantum numbers expected from the quark model, and the model does not predict unobserved states.

There are also higher states of the $q\bar{q}$ system with nonzero orbital angular momentum L. Each given orbital momentum must be represented by the same nine combinations of u, d, and s quarks just discussed for the $L = 0$ case. When the two spins are parallel, their total spin $S = 1$ combines with L to $J = L \pm 1$, and $J = L$. An example where all nine mesons are identified is the $L = 1$, $J = 2^+$ nonet.

One can group the meson states of different L into families which are just rotational excitations of the lightest $L = 0$ meson, with all quantum numbers other than L fixed. These families have some similarity to the rotational "bands" of molecular or nuclear spectroscopy. The data also show that the mass–spin relation is extremely simple: $m^2 \propto J$. The largest well-established rotational family is shown in Fig. 13.

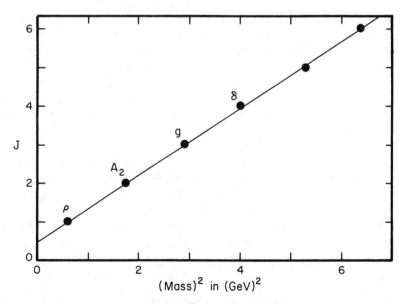

FIG. 13. Rotational excitations of the ρ-meson, showing the linear relationship between J and (mass)^2. Such a plot is called a Regge trajectory. The levels above the g-meson are not fully established.

(c) The necessity of color

The baryons consists of three quarks. We will see again that simple combinations of three quarks correspond to the low-lying observed baryons. There exists a lowest bound state for three quarks which we call $\psi(123)$, where 1, 2, 3 stands for *all* coordinates (position, spin, *and* flavor) of each quark. Here we shall encounter a remarkable situation, because we will only be able to reproduce the observed low-lying levels by making a seemingly contradictory assumption: the baryon ground state $\psi(123)$ must be completely symmetric under the exchange of quarks. This is puzzling since quarks have spin $\frac{1}{2}$, and therefore should have antisymmetric states, obeying the Pauli principle.

The mesons do not present us with this problem since they contain only distinguishable particles; hence the Pauli principle is irrelevant to their spectroscopy. Baryons, on the other hand, contain identical fermions, and then the Pauli principle excludes symmetric states.

Striking evidence for the symmetry of baryon states is provided by baryons containing three identical quarks, such as uuu. The lightest baryon of this composition is $\Delta^{++}(1232)$, with $J = \frac{3}{2}^+$. Several experimental observations show that its wave function is symmetric in the spatial coordinates, and has no nodes. Therefore $L = 0$, and the total angular momentum $J = \frac{3}{2}$ must be entirely due to the spins. This spin state is itself symmetric, giving a wave function that is completely symmetric in all variables, *including* flavor. This example illustrates a large body of data* indicating that *all baryon levels have wave functions that are fully symmetric in space, spin, and flavor*!

We are therefore faced with several drastic alternatives: to discard the quark model; to abandon the Pauli principle; or to ascribe further degrees of freedom to quarks. As the first two are impalatable, we explore the third. This option can be understood by imagining situations that could have arisen had the exclusion principle been discovered before electron spin. The knowledge that the He ground state has two electrons in the same spatial state would have forced the introduction of a two-valued variable for every electron, with which one could then have formed an antisymmetric (opposite spin) wave function in these new variables.

Taking this clue, *we introduce a new discrete variable called color, which is to be ascribed to all quarks, whatever their flavor*. Since Δ^{++} has three quarks in identical space–spin states (and not two, as in the He example), *the exclusion principle requires us to ascribe three possible values to the color variable. Baryon wave functions are then a product of a symmetric space–spin–flavor wave function, and an antisymmetric color wave function.*** If we

---

* The radiative transitions $\gamma N \to \Delta$ provide experimental evidence that the spatial wave functions of $\Delta(1232)$ and the nucleon are the same. Furthermore, elastic eN scattering determines the nucleon charge distribution (see Vol. II, §III.A.1), and does not indicate any zeros. It would also be difficult to construct forces for which the lowest state is not spatially symmetric.

** Just as the Pauli principle must be generalized to include the isospin variable in nuclear physics (recall §D.2), so must the color variable be incorporated in hadron states, which then are antisymmetric under exchange of space, spin, flavor, and color.

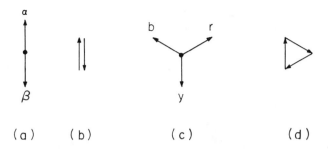

α

β

b   r

y

(a)   (b)   (c)   (d)

Fig. 14. The two possible orientations of a conventional spin (or isospin) are depicted in (a). The state formed when two such spins are combined to form spin zero is symbolized in (b). The possible orientations of a tri-valued "color spin" are shown in (c), while (d) symbolizes the color singlet state, i.e., the baryon state with zero net color.

call the three colors red (r), yellow (y), and blue (b), this color state is

$$\frac{1}{\sqrt{6}}[r_1 y_2 b_3 + y_1 b_2 r_3 + b_1 r_2 y_3 - y_1 r_2 b_3 - r_1 b_2 y_3 - b_1 y_2 r_3], \qquad (5)$$

where, for example, y_2 means that the second quark is in the yellow state.

The three internal colors can also be regarded as the eigenvalues of a tri-valued color "spin" (see Fig. 14). The state (5) is the one in which the color "spins" add to zero, and we may therefore say that it has total color zero. This state is the analogue of the singlet spin state of a two-particle system, where each particle can be in either of the spin states α and β:

$$|0\rangle = \frac{1}{\sqrt{2}}(\alpha_1 \beta_2 - \beta_1 \alpha_2).$$

This singlet has total spin zero and is antisymmetric. The state (5) is also called a color-singlet. The term "singlet" has the following significance. We remember from §D.2 that $|0\rangle$ is invariant under unitary transformations in the space $\mathscr{C}_2$ spanned by α and β, and is therefore a singlet, not a multiplet. The color state (5) is invariant under unitary transformations in the space $\mathscr{C}_3$ spanned by the three independent color states r, y, b, and is therefore also a (color) singlet.* Indeed, the statement "total color zero" expresses just that fact.

There are color states other than (5) that can be constructed for three quarks, but they are not invariant under transformations in $\mathscr{C}_3$. It seems that nature requires three quarks to always be in the singlet state (5); naturally, nonsinglet states of very high mass cannot be excluded.

---

* These and other related statements are proven in Vol. II, §IV.A.

What role does color play in the case of mesons? Since there is no Pauli principle problem, we need not dwell on this question here. Suffice it to say that the color wave function of a $q\bar{q}$-system is

$$\frac{1}{\sqrt{3}} (r_q r_{\bar{q}} + b_q b_{\bar{q}} + y_q y_{\bar{q}}) \qquad (5')$$

where, for example, b_q is the blue state of the quark, and $b_{\bar{q}}$ the blue state of the antiquark. It is also a color singlet, since it is invariant under transformations in* $\mathscr{C}_3$. We may again call it a state of color zero of the $q\bar{q}$ system.

(d) Baryon octets and decuplets

As in the case of mesons, we focus on baryons composed of the light quarks u, d, and s. We expect the lowest-lying baryons to have all constituents in the same orbital $l = 0$ state, so that J should be entirely due to quark spins. As such states are symmetric in space, our hypothesis concerning the symmetry character requires these states to be symmetric under interchange of flavor and spin of any quark pair. We continue to designate this symmetric spin-and-flavor wave function by $\psi(123)$, as for example $\psi(u_\uparrow d_\uparrow s_\downarrow)$, where the arrows indicate the spin direction.

The simplest case is $J = \frac{3}{2}$, in which case all spins are parallel, and the state is symmetric in spin. Consequently it is also symmetric in flavor, so that $\psi(uud) = \psi(udu) = \psi(duu)$, etc. Since there are three flavors to choose from, there are ten distinct $J = \frac{3}{2}$ states of the required symmetry:

$$
\begin{array}{lll}
\psi(u_\uparrow u_\uparrow u_\uparrow) = \Delta^{++} & \psi(u_\uparrow u_\uparrow s_\uparrow) = \Sigma^{*+} & \psi(u_\uparrow s_\uparrow s_\uparrow) = \Xi^{*0} \\
\psi(u_\uparrow u_\uparrow d_\uparrow) = \Delta^{+} & \psi(u_\uparrow d_\uparrow s_\uparrow) = \Sigma^{*0} & \psi(d_\uparrow s_\uparrow s_\uparrow) = \Xi^{*-} \quad (6) \\
\psi(u_\uparrow d_\uparrow d_\uparrow) = \Delta^{0} & \psi(d_\uparrow d_\uparrow s_\uparrow) = \Sigma^{*-} & \psi(s_\uparrow s_\uparrow s_\uparrow) = \Omega^{-} \\
\psi(d_\uparrow d_\uparrow d_\uparrow) = \Delta^{-} & &
\end{array}
$$

Here we use the notation Σ^* and Ξ^* to distinguish these $J = \frac{3}{2}$ states from the lower $J = \frac{1}{2}$ states having the same flavor. Insofar as isospin is concerned, note that the flavor symmetry immediately determines I: for states without s-quarks, we have three isospinors in a symmetric state, so I, like J, must be $\frac{3}{2}(\Delta)$; states with one s-quark are symmetric in the other pair, and therefore have $I = 1(\Sigma^*)$. The remaining three states are obviously an isodoublet Ξ^*, and the singlet Ω.

[We now examine the detailed form of the wave functions ψ in Eq. (6). Let $(1,2,3)$ be a function of the spatial coordinates of three particles 1, 2, 3, which is symmetric under exchange of the particles, for example, the product of three

---

* That the color state $(5')$ is symmetric is related to the symmetric character of the isosinglet state of the $q\bar{q}$ system, $2^{-\frac{1}{2}}(u\bar{u} + d\bar{d})$.

identical functions of the coordinates of 1, 2, and 3. Each particle can have one of the three flavors u, d, or s. We denote by (uds) the state in which particle 1 has flavor u, 2 has flavor d, and 3 has flavor s. In this notation only (uuu), (ddd), and (sss) are also flavor symmetric. So we have $\psi(uuu) = (uuu)$, etc., for these wavefunctions. For other flavor combinations we must symmetrize. For example, for the uud flavor combination, the flavor symmetric form is

$$\psi(uud) = \frac{1}{\sqrt{3}}[(uud) + (udu) + (duu)].$$

The spin state depends on which member of the $J = \frac{3}{2}$ multiplet we consider. In the state $J = \frac{3}{2}$, $m = \frac{3}{2}$, all spins are up: ($\uparrow\uparrow\uparrow$); it is obviously symmetric. In the state $m = \frac{1}{2}$, one of the spins must be down. The corresponding symmetric combination is:

$$\frac{1}{\sqrt{3}}(\uparrow\uparrow\downarrow + \uparrow\downarrow\uparrow + \downarrow\uparrow\uparrow). \tag{7}$$

We get $m = -\frac{1}{2}$ from $m = \frac{1}{2}$, and $m = -\frac{3}{2}$ from $m = \frac{3}{2}$, by reversing all spins. Thus, more explicit expressions for the wave functions of the states (6), can be written as follows for the $m = \frac{3}{2}$ cases with $I_3 \geq 0$:

$$\Delta^{++} = (uuu)(\uparrow\uparrow\uparrow)$$

$$\Delta^{+} = \frac{1}{\sqrt{3}}[(uud) + (udu) + (duu)](\uparrow\uparrow\uparrow)$$

$$\Sigma^{*+} = \frac{1}{\sqrt{3}}[(uus) + (usu) + (suu)](\uparrow\uparrow\uparrow)$$

$$\Sigma^{*0} = \frac{1}{\sqrt{6}}[(uds) + (dus) + (usd) + (dsu) + (sud) + (sdu)](\uparrow\uparrow\uparrow)$$

$$\Xi^{*0} = \frac{1}{\sqrt{3}}[(uss) + (sus) + (ssu)](\uparrow\uparrow\uparrow)$$

$$\Omega^{-} = (sss)(\uparrow\uparrow\uparrow)$$

A product of the form $(uud)(\uparrow\uparrow\uparrow)$ means $(u_\uparrow u_\uparrow d_\uparrow)$ in the notation of Eq. (6); it is the symmetric space function (1,2,3), where particle 1 is a u-quark with spin up, etc. The states with negative I_3 are obtained by exchanging d with u; the states with $m = \frac{1}{2}$ result from replacing ($\uparrow\uparrow\uparrow$) with (7); reversing all spins gives the states with negative m.]

The baryon decuplet (6) is shown as an (I_3, S) diagram in Fig. 15. Because of the complete symmetry between the quarks in these states, it is an equilateral triangle, like the basic quark diagram Fig. 10. The observed baryon spectrum (Fig. 6) shows this decuplet very clearly. Within the decuplet, the $\Delta - \Sigma$, $\Sigma - \Xi$, and $\Xi - \Omega$ mass differences are almost

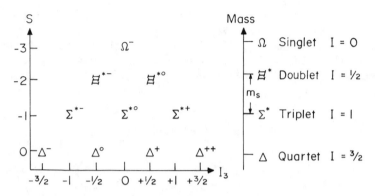

FIG. 15. The baryon decuplet shown in an (I_3, S) diagram. All these states have $J = \frac{3}{2}$, so all quarks have parallel spins. The mass splitting is proportional to S, and is approximately equal to the mass of the s-quark, as shown.

identical (150, 150, and 141 MeV, respectively), and can be understood as being due to the additional (effective) mass of the s-quarks (see Table 3).

What are other possible symmetric three-quark combinations? All three spins need not be parallel; we can have $J = \frac{1}{2}$, which consists of two parallel and one antiparallel spin, $\uparrow\uparrow\downarrow$. This combination is not completely symmetric in the spins, and therefore it cannot be completely symmetric in flavor either. Hence uuu, ddd, and sss are ruled out if $J = \frac{1}{2}$; the asymmetry in spin must be compensated by an asymmetry in flavor if the state is to be completely symmetric.

First we examine the states without s-quarks. There are two possible flavor combinations: $uud(Q = 1)$ and $udd(Q = 0)$, and as $J = \frac{1}{2}$, these are perfect candidates for p and n. Consider uud; to make this symmetric in the u's, we can also choose their spin state as symmetric, so that their spin $J_{uu} = 1$. To this we must add $J_d = \frac{1}{2}$ to give an overall $J = \frac{1}{2}$; the detailed wave function is given below. This state is antisymmetric in the spins of the two possible (u,d) pairs, and also in their flavor, so as to give a totally symmetric state. Obviously the same argument applies to the udd combination. The states with two s-quarks can be analyzed in the same fashion: these two quarks must have parallel spins, and the u or d quark must have its spin opposed to give $J = \frac{1}{2}$. As a result we get two isodoublets, one with $S = 0$, the other with $S = -2$:

$$\psi(u_\uparrow u_\uparrow d_\downarrow) = N^+ = p \qquad \psi(u_\downarrow s_\uparrow s_\uparrow) = \Xi^0$$
$$\psi(d_\uparrow d_\uparrow u_\downarrow) = N^0 = n \qquad \psi(d_\downarrow s_\uparrow s_\uparrow) = \Xi^- \qquad (8a)$$

The remaining $J = \frac{1}{2}$ combinations are uus, uds, dds. There are two possibilities: Either the two nonstrange quarks provide the total isospin

$I = 1$, or $I = 0$. Because of the overall symmetry, the spin of the nonstrange quarks must be $J_{ns} = 1$ if $I = 1$, and $J_{ns} = 0$ if $I = 0$; the spin of s is then added appropriately to give $J = \frac{1}{2}$. $I = 1$ has three substates, $I = 0$ has one; hence, we get four other $J = \frac{1}{2}$ states, one isotriplet and one isosinglet:

$$\psi(u_\uparrow u_\uparrow s_\downarrow) = \Sigma^+$$
$$\psi(u_\uparrow d_\uparrow s_\downarrow) = \Sigma^0 \qquad \psi(u_\uparrow d_\downarrow s_\uparrow) = \Lambda^0 \qquad (8b)$$
$$\psi(d_\uparrow d_\uparrow s_\downarrow) = \Sigma^-$$

[Let us construct the detailed form of the states (8a) and (8b). We start with the proton, $\psi(u_\uparrow u_\uparrow d_\downarrow)$. We again make use of the symmetric function (1,2,3) of the space coordinates of three particles. The two-particle flavor combination $2^{-\frac{1}{2}}(ud - du)$ is antisymmetric; so is the spin combination $2^{-\frac{1}{2}}(\uparrow\downarrow - \downarrow\uparrow)$. Their product is symmetric and has the quantum numbers $J_{ud} = 0$, $I_{ud} = 0$. Adding another u-quark with spin up would give the state

$$\tfrac{1}{2}[(udu) - (duu)](\uparrow\downarrow\uparrow - \downarrow\uparrow\uparrow)$$

with $J = m = \frac{1}{2}$, and $I = I_3 = \frac{1}{2}$, as appropriate to the proton. But this state is not symmetric, because it singles out the third quark. We must symmetrize. For the proton with spin up this yields

$$A\{[(udu) - (duu)](\uparrow\downarrow\uparrow - \downarrow\uparrow\uparrow) + [(uud) - (duu)](\uparrow\uparrow\downarrow - \downarrow\uparrow\uparrow)$$
$$+ [(uud) - (udu)](\uparrow\uparrow\downarrow - \uparrow\downarrow\uparrow)\},$$

where A is a normalization factor. Here a product such as $(udu)(\uparrow\downarrow\uparrow)$ means $(u_\uparrow d_\downarrow u_\uparrow)$, which is the symmetric space function of the quarks as specified. Sorting out the different substates, we find that the $m = \frac{1}{2}$ proton state has the form:*

$$p(m = \tfrac{1}{2}) = \frac{1}{\sqrt{18}} [2(u_\uparrow u_\uparrow d_\downarrow) + 2(d_\downarrow u_\uparrow u_\uparrow) + 2(u_\uparrow d_\downarrow u_\uparrow)$$
$$- (u_\uparrow u_\downarrow d_\uparrow) - (u_\uparrow d_\uparrow u_\downarrow) - (d_\uparrow u_\uparrow u_\downarrow)$$
$$- (u_\downarrow u_\uparrow d_\uparrow) - (u_\downarrow d_\uparrow u_\uparrow) - (d_\uparrow u_\downarrow u_\uparrow)].$$

The neutron ($I_3 = -\frac{1}{2}$) is obtained by exchanging u and d; the $m = -\frac{1}{2}$ states result from reversing all spins.

---

* One arrives at the same expression by choosing another approach: one combines the symmetric two-particle flavor state $uu(I = 1)$ with the symmetric spin state $\uparrow\uparrow$ ($J = 1$). Then one combines this state with a single d-quark, such that the total J and I are $\frac{1}{2}$, and thereafter symmetrizes. This technique was sketched in the paragraph preceding Eq. (8a).

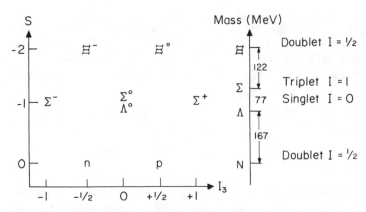

FIG. 16. The baryon octet in an (I_3,S) diagram. All states have $J = \frac{1}{2}$, hence one quark has its spin opposite to that of the others. The mass splittings are indicated on the right.

We get the corresponding functions for other states of the $J = \frac{1}{2}$ octet by replacing some of the ordinary quarks with strange quarks. For example:

$$\begin{pmatrix} \Sigma^+ \\ \Sigma^- \\ \Xi^0 \\ \Xi^- \end{pmatrix} \quad \text{by replacing} \quad \begin{pmatrix} d \text{ by } s \\ u \text{ by } s \\ u \text{ by } s \\ d \text{ by } s \end{pmatrix} \quad \text{in} \quad \begin{pmatrix} p \\ n \\ p \\ n \end{pmatrix}$$

The explicit expressions for Σ^0 and Λ are of a similar form, but somewhat more complicated.⟧

There are altogether eight states listed in Eqs. (8). They comprise the $J = \frac{1}{2}$ *baryon octet*. This octet is shown as an (I_3,S) diagram in Fig. 16. Except for a displacement along the S-axis, and a twofold degeneracy at the center, it looks just like the meson nonet. But in contrast to the mesons, no member of a baryon multiplet $(B = 1)$ is the antiparticle of another member.

As the observed baryon spectrum shows (see Fig. 6), the lowest-lying states fall neatly into such an octet. The mass splittings are indicated in Fig. 16; once again, the mass increases with the number of s-quarks, but the splittings are not equal to each other, as in the decuplet.

To summarize, all low-lying noncharmed baryons fit into two multiplets: the $J = \frac{1}{2}$ octet and $J = \frac{3}{2}$ decuplet. These are precisely the spin–flavor combinations that we expect from the quark model *provided* we assume a totally symmetric space–spin–flavor state. The model predicts that these states have the same intrinsic parity, which also agrees with the data.

A vast array of more highly excited baryon states exist,* as indicated in Fig. 6. Most of those just above the decuplet can be understood as states

---

* See Appendix I, and Close (1979).

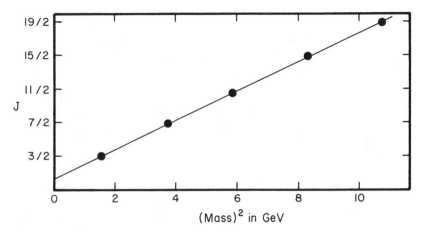

FIG. 17. Rotational excitations of the $J = \frac{3}{2}$, $I = \frac{3}{2}$, state $\Delta(1232)$; see Appendix I.8. As in the case of mesons (see Fig. 13), there is a linear relationship between J and (mass)^2. This is called a Regge trajectory.

where the quarks have nonzero orbital angular momentum. Among these excited states there are also rotational "bands." Some of the best established are shown in Fig. 17; they display the same linear $m^2 - J$ relationship as the rotational "bands" of meson spectroscopy (Fig. 13).

(e) Mesons composed of heavy quark–antiquark pairs

The hadronic spectra we have discussed thus far can be understood in a qualitative manner in terms of the quark model: quantum numbers, level orderings, and multiplet structures, can be accounted for. But it is not yet possible to compute their energies and spatial wave functions without ad hoc models. There are, however, two hadronic families whose spectra have the same character as those of hydrogenic atoms, and which can be understood quantitatively. In so doing one also gains considerable insight into the forces between quarks, that is, into the primordial strong interaction.

The hadrons in question are mesons built of heavy quark–antiquark pairs, a topic that we shall treat in detail in §III.B, Vol. II. Figure 8 shows two families of this type: the $c\bar{c}$ states η_c, J/ψ, χ, η_c', ψ', etc., and $b\bar{b}$ states Υ, Υ' etc. Their characteristic excitation energies are of order several hundred MeV, as in the light hadrons, but this is small compared to the masses of the ground states (~ 3 GeV for $c\bar{c}$, ~ 10 GeV for $b\bar{b}$). Consequently, these are nonrelativistic systems,* and that is why they are amenable to an analysis that is both simple and quantitative.

* This nonrelativistic nature can also be recognized from the smallness of the splittings due to $q\bar{q}$ spins: η_c and J/ψ are the spin singlet (0^-) and spin triplet (1^-) $c\bar{c}$ ground states, separated only by 100 MeV, in contrast to the $\pi - \rho$ and $K - K^*$ splittings of ~ 600 MeV and ~ 400 MeV, respectively. In atomic physics spin splittings are due to magnetic interaction, and are of order $(v/c)^2$. This is also believed to be the case with strong interactions.

If the motion is nonrelativistic the system can be described by a Schrödinger equation containing a static $q\bar{q}$ potential V. A simple potential that does this with very good accuracy, and which even has a certain measure of theoretical motivation, has been found (see §III.B.3 and §IV.D, Vol. II). This potential is plotted in Fig. 18; as the $q\bar{q}$ separation r tends to zero, it

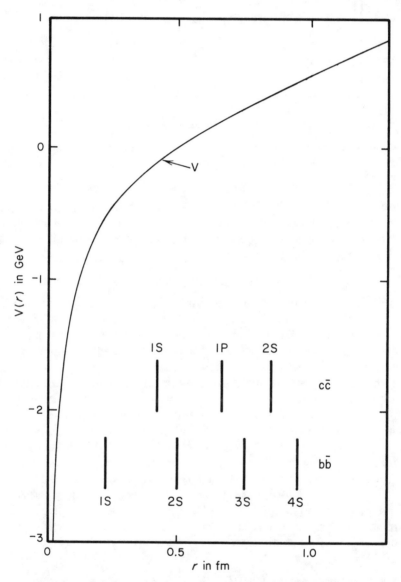

FIG. 18. The potential $V(r)$ that represents the interaction of a heavy quark–antiquark pair, in particular $c\bar{c}$ and $b\bar{b}$. The analytic form of the curve shown is motivated in §IV.C.4 (Vol. II). The vertical bars show the rms radii of various $c\bar{c}$ and $b\bar{b}$ wave functions computed from this potential.

behaves approximately like a Coulomb potential $(1/r)$, but at large r it rises linearly, and traps the $q\bar{q}$ pair, no matter what the energy. This confining property of V is supposed to account for the observation that quarks cannot be liberated from hadrons. The linear rise with r gives a constant force at large separation, and as we shall learn in §IV. D.4 (Vol. II), this leads to the linear $m^2 - J$ relationship shown in Figs. 13 and 17.

The spectrum computed from the potential of Fig. 18 is shown in Fig. 19. In the case of the ψ-family, all levels up to and including the $J = 1$ D-state have been observed. The correspondence between the states of Figs. 8 and 19 is as follows: $\eta_c = 1^1S$, $\eta'_c = 2^1S$, $\psi = 1^3S$, $\psi' = 2^3S$, and $\chi = 1^3P$. In the Y family the lowest four ^3S states and the first two ^3P states are established at this time. The agreement between the spacings in the theoretical level scheme, and the observed mass spectra, is very good (see §III.B.3, Vol. II). It is remarkable that the same potential reproduces both the $c\bar{c}$ and $b\bar{b}$ spectra, although the c and b masses differ by a factor of 3. This is one of the most clear-cut pieces of evidence that the strong forces between quarks do not depend on flavor (recall §5(d)).

A considerable number of radiative dipole transitions between members of the ψ-family, and the Y-family, have been observed (Fig. 19 and Appendix I). The strengths of these transitions agree reasonably well with those computed from the wave functions associated with the potential V. The cross section for production of the ψ and Y levels in e^+e^- annihilation can also be computed with considerable success from these same wave functions.

7. The strong interaction field: quantum chromodynamics

(a) General considerations

If quarks are the basic hadronic building blocks, the structure of hadrons, and all "strong" processes involving them, must, ultimately, be due to interactions between quarks. The strong interactions must have the following properties in order to account for the observed facts, many of which were discussed in the three preceding sections:

   I. The strong interaction between quarks does not depend on their flavor.

   II. The strong interaction conserves the quantum numbers I_3, S, and C, as well as the flavor quantum number carried by b, the heaviest quark now known. If still heavier quarks exist, they presumably carry new conserved flavor quantum numbers.

   III. The forces between quarks must make it very difficult—perhaps impossible—to separate a quark (or antiquark) from a hadron, and they must be such that at presently available energies the only observed states have the composition $q\bar{q}$, qqq, or combinations thereof, in color states given by Eqs. (5) and (5').

   IV. For massive slowly moving quarks, the forces are represented by a potential having the form shown in Fig. 18.

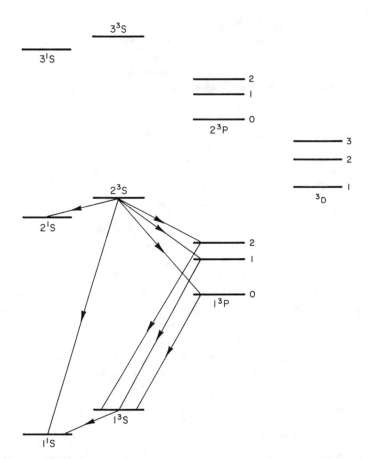

FIG. 19. The low-lying excitation spectrum for a nonrelativistic quark–antiquark pair as inferred from the potential of Fig. 18. Nonrelativistic quantum mechanics gives a spectrum that only depends on the relative angular momentum L of the pair. Because of the analogy between QCD and QED (see §7 below), one expects that the strong interaction has forces analogous to the magnetic interactions which produce spin splittings in atoms (fine and hyperfine structure, see Vol. II, §II.C.6). A state with a given L then splits into three levels with $J = L - 1$, L, and $L + 1$ when the total spin $S = 1$ (spin triplet); there is also another state with $J = L$ when $S = 0$ (spin singlet). The states are designated by $^{2S+1}L_J$, where S, P, D, etc. stand for $L = 0, 1, 2, \ldots$, respectively. For $L > 0$, J is indicated for each level. In S states, where $L = 0$, there is no need to specify J since $J = 1$ for triplets and 0 for singlets.

The gross structure of the spectrum does not change significantly in going from the $c\bar{c}$ to the $b\bar{b}$ case (roughly a tripling of quark mass), but the spin splittings change appreciably. The splittings in this figure are enlarged for the sake of clarity; the observed $c\bar{c}$ spectrum is shown in detail in Vol. II, Fig. III.22; our knowledge of these splittings in the $b\bar{b}$ system is still quite incomplete. Only S-state spin singlets are shown, because it will be very difficult to find singlets with $L > 0$.

As we shall learn in §8(c), the ^3S and $^3D_1$ states are directly visible in e^+e^- annihilation. By this technique the first half dozen ^3S states, and the $^3D_1$ state shown here, have all been identified in the ψ family, while the lowest four ^3S states have been found in the Υ-family. All the indicated photon transitions in this figure have been detected in the ψ family. For further details, see Appendix I.6 and §III.B(Vol. II).

What is the nature of the strong interaction? Gravitational and electromagnetic interactions are successfully described in terms of fields that are produced by sources—energy in the first instance, charge in the second. Moreover, both of these interactions can be approximated by a simple potential (Newtonian or Coulombian) for nonrelativistic motions, but require the introduction of a field when the particles move with velocities comparable to the speed of light. It is tempting to describe the strong interactions in a similar way. We therefore make the following premise: *There exists a strong-interaction field, which has its own degrees of freedom, and can propagate by itself. The quarks are sources and sinks of this field, and therefore the field mediates interactions between quarks.*

What attribute of a quark plays the role that the charge plays in the electromagnetic case? It should determine the strong field emanating from the quark, and its interaction with an externally applied strong field due, say, to a passing quark in another hadron. Clearly this attribute cannot be flavor or electric charge, for that would contradict (I) above. On the other hand, the baryon spectrum tells us that qqq states antisymmetric in color lie far (perhaps infinitely far!) below those having a different color configuration. This suggests the hypothesis *that the strong interaction field is coupled only to the quark color.* By this we mean that the field is the same for all flavors, and that a quark does not change its flavor when the field is emitted or absorbed.

As we have seen in §6, all observed hadron states are color singlets; they are invariant under unitary transformations among the three quark colors. Hence our earlier statement (III) should be rephrased as follows: *the interaction between quarks and the strong field must produce a strong bond in color singlets, whereas color states other than singlets must acquire a high (possibly infinite) mass.* Among other things, this would account for the difficulty (impossibility?) of isolating quarks, because they are members of a noninvariant color triplet. The existence of a free quark would then require a high (perhaps infinite) energy.

The recognition that color-invariant states are of such singular importance suggests that color invariance is a symmetry of nature. The field theory based on this symmetry, and on color as the source of fields, is called *quantum chromodynamics* (QCD). It is described in some detail in Vol. II, Chap. IV. In this volume we will restrict ourselves to brief descriptions of the most salient features of QCD.

(b) Quantum chromodynamics

In many respects QCD is similar to quantum electrodynamics (QED). The strong field is also a vector field having "color-electric" and "color-magnetic" components. The field quanta are called *gluons*. These correspondences between QED and QCD can be summarized as follows:

QED	QCD
electron	quark
charge	color
photon	gluon
positronium (e^+e^-)	mesons $(q_1\bar{q}_2)$

The profound difference is that the QCD source has a trivalued quantum number, and is not a scalar as in QED. Emission and absorption of gluons is accompanied by changes in color. Color symmetry requires that the total color is conserved, and therefore it follows that *the field must carry color, too*! Field theories in which the "charge" can be transferred to the field are called "non-Abelian" theories.

Figure 20 shows the fundamental vertices of QCD, just as Eq. C(12) gives the fundamental vertex of QED. Diagram (a) describes the transitions between a quark state and a quark-plus-gluon state or, if read from left to right, quark–antiquark annihilation into a gluon. At first sight this would

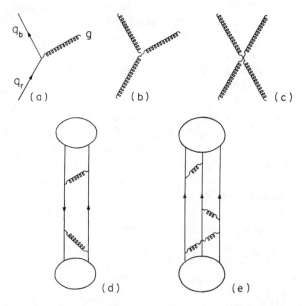

FIG. 20. The fundamental Feynman diagrams of QCD. The quark–gluon, three-gluon, and four-gluon vertices are shown in (a), (b), and (c), respectively, where the solid straight line represents a quark, and coiled lines gluons. The first vertex has its analogue in QED, [see Eq. C(12)], but the others are novel to QCD. They stem from the intrinsically nonlinear nature of QCD, which is due to the fact that the gluon carries color. Thus (b) represents the bremsstrahlung of a gluon by a gluon, while (c) gives gluon–gluon scattering.

The binding of quarks into mesons and baryons is depicted by diagrams (d) and (e) within the framework of QCD. As these are long-lived bound states, an arbitrary number of gluons are actually exchanged between the quarks, and by virtue of the vertices (b) and (c), these gluons can interact with each other.

lead one to suppose that there are nine distinct gluons, corresponding to the nine combinations $r\bar{r}$, $r\bar{b}$, $r\bar{y}$, $b\bar{r}$, ... , $y\bar{y}$. But as already explained in connection with Eq. (5′), the symmetric combination ($r\bar{r} + b\bar{b} + y\bar{y}$) carries no color, and it is therefore not coupled to the gluon field. Consequently *there are only eight independent color states of the gluon field.*

Since gluons are coupled to all colored objects, and are themselves colored, they are coupled to themselves. For that reason QCD is an intrinsically nonlinear field theory. This is to be contrasted with QED, where the field quantum—the photon—carries no charge. (QCD actually bears a similarity to Einstein's intrinsically nonlinear theory of gravitation, where the gravitational field is coupled to all forms of energy, including its own.) Hence QCD has vertices that have no QED counterpart: the emission or absorption of a gluon by a gluon, Fig. 20(b), and direct gluon-gluon scattering, Fig. 20(c).

There are compelling arguments, though no rigorous mathematical proof, that the nonlinear nature of QCD is responsible for the confinement of color. The mechanism is most readily described in the case of a quark–antiquark system. If we forget the nonlinearity for a moment, a $q\bar{q}$ system would be like an electromagnetic dipole, with its familiar pattern of spatially dispersed field lines, as shown in Fig. 21(a). The nonlinear character of QCD produces a compression of this pattern into a so-called *color flux tube*, as if there were an attractive force between the field lines, giving the result shown in Fig. 21(b). As the distance r between the pair is increased, the cross-sectional area A of the tube remains constant. But the number of field lines only depends on the color of the sources (just as the charge determines the number of electric field lines), hence the field strength in the tube also remains constant, and the field energy of the configuration grows in proportion to the volume Ar of the tube.

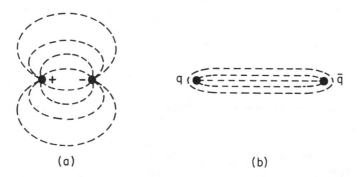

(a)   (b)

FIG. 21. The electric field surrounding two opposite charges is shown in (a), and the color field surrounding a $q\bar{q}$ pair is shown in (b). In QCD the lines of force are compressed into a "flux tube," as if there were an attraction between the lines of force. As the $q\bar{q}$ separation r is increased, the cross section of the tube remains constant. As explained in the text, the energy therefore grows in proportion with r, and gives the linear asymptote shown in Fig. 18. In reality, the flux tube cannot be stretched forever, because it becomes energetically favorable to create a quark pair out of the vacuum, which then results in a transition from a one-hadron to a two-hadron state.

Since the area A is fixed, the energy of the $q\bar{q}$ system increases linearly with its separation r, and no process that transfers a finite amount of energy to the system can separate a quark or antiquark.

The same consideration applies to any composite color singlet: If one separates a colored portion of that system (say a qq pair in a baryon) from the remainder, the energy grows linearly and confines the colored constituents. Since gluons also carry color, they, too, cannot exist in isolation. Like quarks, they only appear inside hadrons.* On the other hand, this confinement mechanism does not prevent the separation of a hadron into two hadrons, as in $\Delta \to N\pi$, because all hadrons have vanishing net color. How such processes are described by QCD will be discussed in the next subsection.

A further result of QCD concerns the potential energy V of a non-relativistic quark–antiquark pair. It appears that V actually has the form shown in Fig. 18, but this is only rigorously established at short distances.** The linear asymptote just discussed, and shown in Fig. 18, is consistent with the data on heavy quark–antiquark spectra, and most particularly, with the rotational spectra, as exemplified by Figs. 13 and 17.

Finally, one may ask how QCD accounts for isospin conservation. Flavor conservation is, of course, guaranteed by the basic hypothesis that in emitting or absorbing gluons, quarks may change their color, but not their flavor. But flavor conservation only assures the constancy of I_3, not of I. Conservation of I requires, in addition, that the (u,d) pair be degenerate in mass or, at least, that their mass splitting be negligible in comparison to typical hadronic masses (several MeV or less). Unfortunately, QCD has nothing whatsoever to say about the quark mass spectrum, nor, for that matter, does any other existing theory. Consequently we have the rather embarrassing situation that isospin conservation, which plays so important a role in hadronic phenomena, is the result of a degeneracy that is "accidental" within the current framework. By the same token, the approximate flavor $SU(3)$ symmetry exploited in Figs. 10, 11, 15, and 16 is also an "accident" of the quark mass spectrum—the relatively small mass difference between s and (u,d).

(c) Hadronic processes in quantum chromodynamics

We shall briefly describe how the quark–gluon theory depicts some of the most familiar processes involving hadrons by means of the fundamental Feynman diagrams of Fig. 20. The quark–gluon vertex (diagram a) underlies all quark–quark interactions. Indeed, mesons and baryons are $q\bar{q}$ and qqq states bound by the exchange of gluons, as shown in Figs. 20(d) and (e). In these diagrams depicting hadrons the number of gluon lines is of no significance, since gluons are exchanged continuously within hadrons.

As a simple example of an actual process, consider the emission of a pion by a hadron. We choose Δ^{++}, whose composition is uuu. Figure 22(a) shows

---

* Two or more gluons can be combined into color singlet states, and should appear in the hadron excitation spectrum. At this time no such "glue-balls" have been positively identified.
** QCD has the rigorously established property of "asymptotic freedom": The force between sources of the color field is logarithmically weaker than $1/r^2$ as the separation r tends to zero.

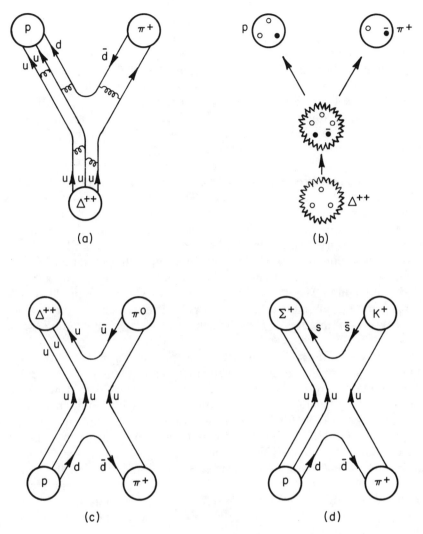

FIG. 22. Diagrams depicting hadronic processes in QCD. The decay $\Delta^{++} \to p\pi^+$ is shown as a Feynman diagram in (a), and in a more naive manner in (b). The two-body reactions $p\pi^+ \to \Delta^{++}\pi^0$ and $p\pi^+ \to \Sigma^+ K^+$ are shown in (c) and (d), respectively.

$\Delta^{++} \to p\pi^+$. One of the gluons inside the Δ^{++} produces a $d\bar{d}$ pair, and the $\bar{d}$ unites with a u to become π^+. Another way of picturing this process is shown in Fig. 22(b). This illustrates how a hadron can transform some of its color field into quark–antiquark pairs and, if the energy suffices, a $q\bar{q}$ pair then forms a color singlet and leaves as a meson.

Another example is the reaction $p\pi^+ \to \Delta^{++}\pi^0$, shown in Fig. 22(c). (From now on we leave out the coiled lines representing gluons.) When the π^+ and the p merge, the d within the proton annihilates with the $\bar{d}$ within the π^+, and a $u\bar{u}$ pair is created. The three u's combine to the Δ^{++}, and the remaining $u\bar{u}$ pair form π^0. Associate production, e.g., $p\pi^+ \to \Sigma^+ K^+$, is shown in Fig. 22(d); it is quite similar to the previous reaction, except that an $s\bar{s}$-pair is created.

The nuclear force must also be encompassed by QCD. At distances large compared to the size of hadrons there is no strong force between hadrons since they are color neutral. This is analogous to the absence of a force between neutral atoms at large distances since they are electrically neutral. When two or more atoms are at distances comparable to their size, the mutual interactions cause some changes in their internal structure. These distortions give rise to relatively short-ranged forces between the atoms which are known as Van der Waals forces, or chemical forces. The same is expected to happen if two nucleons come near each other. The mutual interactions of the two three-quark systems give rise to an attraction; loosely speaking, it is the Van der Waals force between two three-particle systems. At this time there is no quantitative theory of these forces, because one still does not know how to solve the field equations of QCD in such a complicated situation.*

We now consider a different class of processes: the creation of hadrons in e^+e^- annihilation. These reactions proceed in two steps: the production of a $q\bar{q}$ pair, and the subsequent fate of that pair. The first step will be treated in the next section since it is an electromagnetic process. What about the second step?

Were it not for the strong interaction, the primary $q\bar{q}$ pair would escape to infinity, as shown in Fig. 23(a) where the pair happens to be $s\bar{s}$. But the quark-confining strong interaction between the primary pair grows without bound as the $q\bar{q}$ separation increases, and makes this impossible. Instead, it is energetically favorable for the strong field to produce a secondary quark pair $q_1\bar{q}_1[u\bar{u}$ in Fig. 23(b)]. Then q_1 bonds to $\bar{q}$, and $\bar{q}_1$ to q, to form color singlet mesons m and $\bar{m}$. There are no long-range confining forces between color singlets, and therefore the process $e^+e^- \to m\bar{m}$ can actually occur, as shown in Fig. 23(b) for the case $e^+e^- \to K^+K^-$. More mesons can be produced at higher energy if m or $\bar{m}$ is formed in an excited state m^*, for that will then decay, as in $e^+e^- \to K^*\bar{K} \to K\pi\bar{K}$, where the decay $K^* \to K\pi$ is shown in Fig. 23(c).

(d) Quark and gluon jets

When the energy W of the e^+e^- pair is very large compared to meson masses, the argument of the preceding paragraph leads to striking conclusions.

---

* Some readers may be familiar with the Yukawa theory of the nuclear force. In that theory the nucleon–nucleon interaction is due to the exchange of pions, in analogy to the photon-exchange description of the Coulomb force. In the Yukawa theory nucleons and pions are assumed to be point-like elementary particles. Since they are not, the pion-exchange description of the nuclear force is, at best, an approximation. The relationship between the quark–gluon picture of a nucleon–nucleon interaction, and the meson-exchange picture, is shown in Fig. 24. It turns out that the one-pion exchange interaction does describe the asymptote of the nucleon–nucleon force when the internucleon distance is of order one fermi or more.

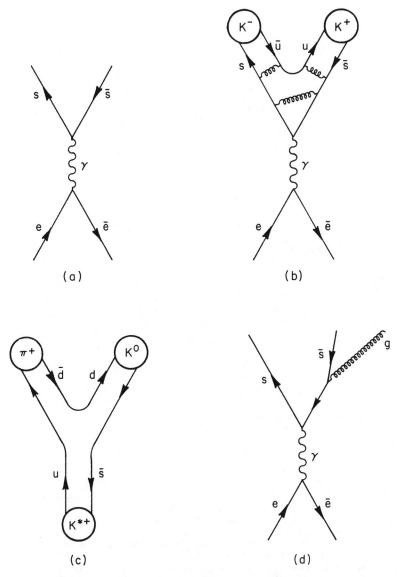

FIG. 23. e^+e^- annihilation into hadrons is a two-step process in QCD. First a quark–antiquark pair is created by a virtual photon, as in (a). Then the color-confining processes discussed in the text come into play and produce further $q\bar{q}$ pairs which combine with the primordial pair to form hadrons, as in (b), which shows a two-meson state. The mechanism that converts the $s\bar{s}$ pair into $K\bar{K}$ in (b) is also responsible for purely hadronic decays such as the one shown in (c). Gluon bremsstrahlung by a very energetic quark is shown in (d), but the subsequent conversion of this hard gluon into a hadron jet is not depicted here. For experimental results relevant to this diagram see Fig. 27.

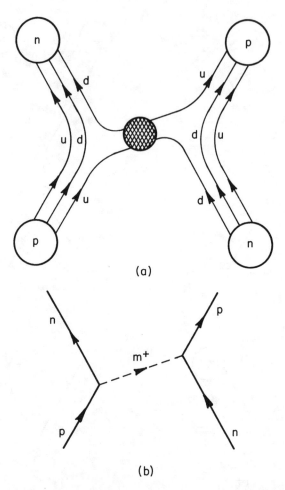

(a)

(b)

FIG. 24. The nuclear force due to one-meson exchange. In (a) a quark level diagram contributing to np scattering is shown. Obviously, this is but one diagram out of an infinite set, but it has a
special significance in that the exchanged $u\bar{d}$ pair has the same quantum numbers as the lightest of
all hadrons, the pion. This interacting $u\bar{d}$ pair, represented in (a) by a shaded circle, is in a state
that is a coherent superposition of all $I = I_3 = 1$ mesons m^+ (i.e., π^+, ρ^+, A_1^+, A_2^+, etc.) It can
be shown that, at low energy, (a) can be approximated by a sum of terms of the type (b), wherein
the solid lines represent elementary point-like fermions, and the dashed line elementary
point-like bosons. Each diagram of type (b) is therefore equivalent to a force, and the range of
this force is given by the Compton wavelength of m^+. For that reason the lightest object
dominates at large nucleon–nucleon separation. Within the framework of QCD, this is the origin
of the Yukawa one-pion exchange theory of the nuclear force.

In Fig. 25 we visualize the course of events after the high energy process
$e^+e^- \rightarrow \gamma \rightarrow q\bar{q}$ has occurred. In the c.o.m. frame (which is the laboratory
in the case of colliders), the primary quarks move along $\hat{n}$ in opposite
directions with large momenta [see Fig. 25(b)]. As we have just learned, the
field will then cause a secondary pair to be produced. In contrast to the

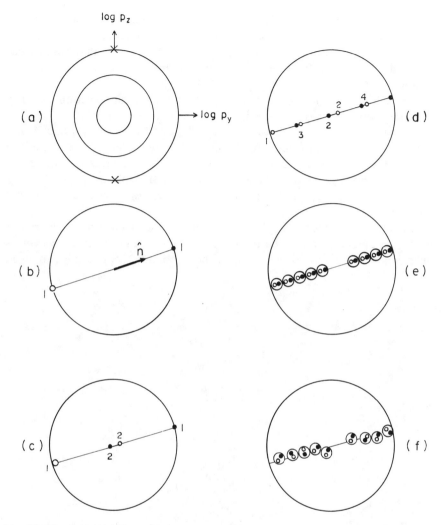

FIG. 25. The temporal development of a quark–antiquark jet. These plots are to be thought of as snapshots in a logarithmic momentum space. In (a), the concentric circles correspond to momenta $W/2$, $W/20$, and $W/200$, where $W/2$ is the incident momentum (or energy—we are in the ultrarelativistic regime). A linear p scale would actually be very obscure because the momenta of particles produced differ by large factors; the choice of $\log p$ overcomes this, and also has a theoretical motivation (it is essentially the rapidity, as defined in Vol. II, Appendix III).

The first frame (a) shows the $e\bar{e}$ pair marked as crosses. The next, (b), the primary pair $q_1\bar{q}_1$, where q's are solid dots and $\bar{q}$'s open circles. Diagrams (c) and (d) show intermediate stages where first one, and then two more pairs are created out of the vacuum. Frame (e) is the final state in this case, with four mesons in the right hemisphere and five in the left. Finally, (f) depicts the final state more realistically by including small transverse momenta. (Note that the primordial pair is close to the outer circle because of energy conservation, and the logarithmic momentum scale.)

105

low-energy situation, a primary and secondary quark cannot bond into stable mesons because of the large relative momenta between the primaries and secondaries [see Fig. 25(c)]. Pair production must therefore occur repeatedly until there are enough $q\bar{q}$ pairs with low relative momenta to form a stable multihadron state [see Figs. 25(d)–(f)].

We shall make one assumption concerning this chain reaction: when a secondary quark pair is created due to the spatial separation of a pair with large relative momenta, the total energy E_{sec} of the newly produced pair in the c.o.m. frame of the producing pair is of order a typical hadron mass (≤ 1 GeV). In the laboratory frame such secondaries can therefore have large momenta along the direction of motion of the primordial $q\bar{q}$ pair if $W \gg E_{sec}$, but only momenta of order E_{sec} transverse to that direction.

This assertion concerning the total energy of secondary quark pairs has three consequences: First, the number of mesons produced rises forever as the primary e^+e^- energy increases without bound; second, the produced

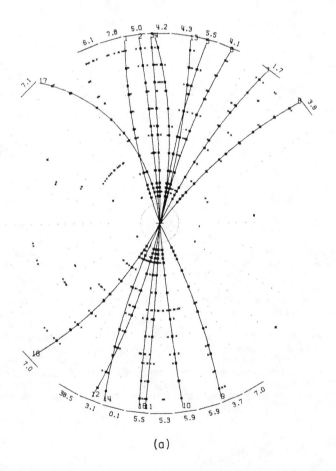

(a)

mesons have momenta that lie within narrow cones surrounding the direction of the primordial q and $\bar{q}$; third, the probability that such *jets* have the orientation $\hat{n}$ relative to the e^+e^- direction is determined by the angular distribution of the primordial process $e^+e^- \rightarrow q\bar{q}$. The electromagnetic mechanism responsible for $e^+e^- \rightarrow q\bar{q}$ is the same as that for $e^+e^- \rightarrow \mu^+\mu^-$, where μ is the muon, a heavy replica of the electron, which will be discussed in detail in §9. Therefore the jet axis angular distribution should coincide with that for $e^+e^- \rightarrow \mu^+\mu^-$.

The data confirm these expectations. Indeed, at high energies $(W \gtrsim 10 \text{ GeV})$ the majority of e^+e^- final states is observed to have a two-jet structure. Figure 26(a) shows such an event; it gives an almost visual demonstration of the ejection of quarks. (The quarks are not actually seen, of course, just the mesons produced in their wake.) The angular distribution of two-jet events is identical to that for $e^+e^- \rightarrow \mu^+\mu^-$, as shown in Fig. 26(b). This is to be expected since, in both cases, a fermion–antifermion pair is created electromagnetically.

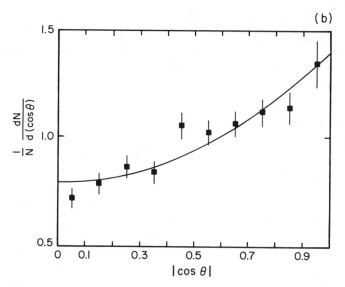

FIG. 26. (a) A two-jet event from $e\bar{e}$ annihilation at a c.o.m. energy of 35 GeV as observed at DESY (TASSO Collaboration, 1980). The event is viewed along the direction of the $e\bar{e}$ beam line. The detector is in a magnetic field, so that the tracks of charged particles are bent, permitting a determination of their momentum, and the sign of their charge. This instrument only detects particles on a discrete mesh defined by a set of wires. The tracks shown are interpolated by computer.

(b) The angular distribution of two-jet events, as measured by the CLEO detector at the Cornell Electron Storage Ring (Cabenda, 1982). For each two-jet event of the type shown in (a), a jet axis is defined, and the angle between this axis and the incident $e\bar{e}$ direction is measured. This procedure yields the points shown. The solid curve is the angular distribution computed with pure QED for the process $e\bar{e} \rightarrow q\bar{q}$, or equivalently, for the process $e\bar{e} \rightarrow \mu\bar{\mu}$ (see §II.C.2, Vol. II).

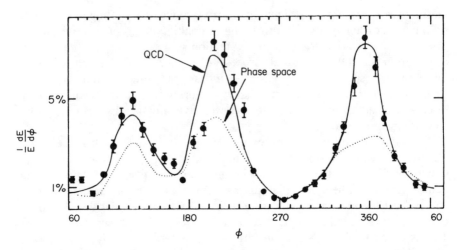

FIG. 27. Evidence for gluon jets. At the highest c.o.m. energies presently available in $e\bar{e}$ annihilation, there is a class of events which does not have the two-jet form of Fig. 26(a). These events have a planar configuration. In this plot the energy transported outward by the produced hadrons is shown as a function of the azimuthal angle ϕ in the event plane. The data, from the MARK J Collaboration (1980) at DESY, show three distinct jets. Presumably two are due to a $q\bar{q}$ pair, the third to a gluon. The angular distribution expected from the Feynman diagram of Fig. 23(d) is shown as the solid curve marked QCD, while the dotted curve marked "phase space" is the prediction of a model which has no correlations between particles beyond the demands of energy and momentum conservation.

By virtue of the basic quark–gluon interaction, one (or more) gluons may be radiated when the primary quark pair was ejected. Instead of the more frequent $\gamma \rightarrow q\bar{q}$ process, we then would have $\gamma \rightarrow q\bar{q}g$, where g stands for a gluon [see Fig. 23(d)]. This is analogous to the bremsstrahlung process $\gamma \rightarrow e^+e^-\gamma'$ that accompanies pair creation, $\gamma \rightarrow e^+e^-$. The gluon also interacts strongly with the quarks and upon separation, the increasing field between gluons and quarks will produce secondary quark pairs with low transverse momenta. If all three particles $q\bar{q}g$ have high momenta, and appreciable angular separation from each other, one expects to see three jets emerging. Events of this character are now accumulating, and should eventually provide direct evidence for the existence of gluons (see Fig. 27).

Jet formation is not restricted to e^+e^- annihilation; high-energy hadron–hadron collisions, and highly inelastic scattering of electrons, muons, and neutrinos by nucleons can impart very large momentum transfers to individual quarks. Such a quark is then ejected from its parent hadron, and the strong force between it and the other quarks will bring the jet production mechanism into play. Indeed, events having a jet-like structure have been observed in all the processes listed in the first sentence of this paragraph (see Fig. 28).

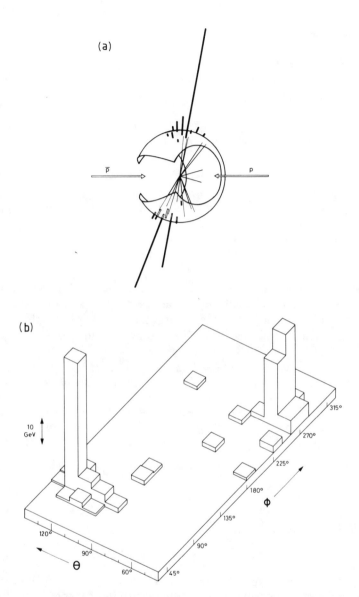

(a)

$\bar{p}$　　p

(b)

10
GeV

315°

270°

225°

180°

Φ

135°

120°

90°

90°

θ

60°

45°

FIG. 28. A two-jet event in a $p\bar{p}$ collision having a total c.o.m. energy of 540 GeV, as observed in the UA2 Detector at CERN (Banner, 1982). This detector consists of an inner spherical counter that is sensitive to charged particles, surrounded by a "calorimeter," a device that measures all energy deposited within a specified element of solid angle by charged *and* neutral particles. In (a) the lines emanating from the $p\bar{p}$ collision point are the tracks of charged particles, while the heavy lines have a length proportional to the energy deposited in the indicated angular cell of the calorimeter. This angular distribution is shown in more detail in (b), where the histogram shows the energy in each cell as a function of the polar and azymuthal angles, θ and ϕ. The produced particles are in two rather narrow cones, and these cones contain a sizeable portion of the available energy.

8. The electromagnetic interaction of hadrons

(a) Conservation laws

The interaction of hadrons with the electromagnetic field has already been mentioned on several occasions. In §5(d) we learned that this interaction conserves all flavors:

$$\Delta I_3 = 0, \ \Delta S = 0, \ \Delta C = 0. \tag{9a}$$

Conservation of I_3 does not tell us whether I is conserved. As we shall see, the electromagnetic interaction can change I. But this change is restricted: in emitting or absorbing one photon, the total isospin of a hadron can change by at most one unit:

$$\Delta I = 0 \text{ or } 1. \tag{9b}$$

Examples of $\Delta I = 0$ transitions are $D^* \to D\gamma$ and $K^* \to K\gamma$, while $\Sigma^0 \to \Lambda\gamma$ and $\gamma N \to \Delta$ are $\Delta I = 1$ transitions. The absence of transitions with $\Delta I > 1$ has been verified in nuclear physics. States that differ by more than one unit are not found in simple hadrons. They occur only in nuclei.

⟦The selection rule (9b) follows from the nature of the electromagnetic interaction, H_{em}. Insofar as hadrons are concerned, the basic processes* are the conversion of a field quantum into a quark pair $\gamma \to q_i \bar{q}_j$, and processes related to it by crossing, such as $q_i \to q_j \gamma$. Because of (9a) the flavor indices must be the same, $i = j$. In general, γ converts into a linear superposition of flavor neutral $q\bar{q}$ pairs, with relative amplitudes given by the quark charges Q_i:

$$\gamma \to e \sum_i Q_i |q_i \bar{q}_i\rangle = e(\tfrac{2}{3}|u\bar{u}\rangle - \tfrac{1}{3}|d\bar{d}\rangle + \cdots). \tag{10}$$

For now we disregard the pairs with other flavors since they cannot contribute to the isospin. Equation (10) can be written as

$$\gamma \to \tfrac{1}{6}(|u\bar{u}\rangle + |d\bar{d}\rangle) + \tfrac{1}{2}e(|u\bar{u}\rangle - |d\bar{d}\rangle). \tag{11}$$

Here the first and second terms are the $I = 0$ and $I = 1$ $q\bar{q}$ combinations respectively, as mentioned in §6(b). Any further development of (11) due to the strong interaction will not change the isospin, which proves that γ (having no isospin) can only convert into hadronic states having $I = 0$ or $I = 1$.

We shall now show that (11) imposes selection rules on the radiative transition $H \to H' \gamma$, where H and H' are hadronic systems. Consider the case where the $q\bar{q}$ state evolves, via strong interaction processes, into $\bar{H}H'$, where $\bar{H}$

---

* Because of energy and momentum conservation these basic processes cannot occur with free particles. However, any electromagnetic process is a combination of these processes, with one or more of the participating particles in an intermediate state, where energy conservation does not hold.

is the antisystem to H:

$$\gamma \to \tfrac{2}{3}e|u\bar{u}\rangle - \tfrac{1}{3}e|d\bar{d}\rangle \to e|H\bar{H}'\rangle. \tag{11'}$$

By crossing, this gives the radiative transition $H \to H'\gamma$. We now take advantage of two facts: (a) H and $\bar{H}$ have the same total isospin; (b) the difference in isospin of $\bar{H}$ and H' must be either 0 or 1 since (11') requires $\bar{H}H'$ to have $I = 0$ or 1. In the general radiative transition $H \to H'\gamma$ the change in I is therefore 0 or 1, which establishes (9b).]

(b) Radiative transitions and magnetic moments

Most of the known radiative transitions are dipole transitions. Because the internal motions of hadrons composed of light quarks are relativistic, higher multipoles are not as inhibited as in atomic and nuclear physics. But the emission of a higher multipole requires a correspondingly larger change of spin, and this inevitably requires a rather highly excited emitter. The latter can de-excite by meson emission, which swamps the electromagnetic decays.*

The best studied $E1$ transitions occur in the ψ-family: $\psi' \to \chi\gamma$ and $\chi \to \psi\gamma$, where ψ' and ψ are S-states, while χ is a P-multiplet [see Figs. 8(b) and 19]. An $E1$ amplitude is proportional to a matrix element of the electric dipole moment, and therefore measures both the quark charge and the radial structure of the wave functions. Numerous $E1$ transitions between members of the Y-family are also expected, and some have been observed at the rate expected by theory.

The electromagnetic interaction also causes multiphoton radiative transitions. A striking example is the decay $\pi^0 \to 2\gamma$. Recall that π^0 is composed of $u\bar{u}$ and $d\bar{d}$ pairs (see Table 4). These can annihilate by the mechanism responsible for $e^+e^- \to 2\gamma$ shown in Eq. C(13), i.e., by the diagram

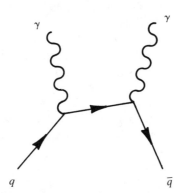

---

* This is illustrated by $A_2(1318) \to \pi\gamma$, a magnetic quadrupole transition with a branching fraction of only $\tfrac{1}{2}\%$ because the $J = 2^+$ level A_2 has a width of ~ 100 MeV due to a variety of hadronic decays ($\rho\pi, \eta\pi, \omega\pi\pi$, etc.). In heavy quark mesons (especially $b\bar{b}$) states exist with $L \geq 2$ that are very stable against hadron emission; but these are nonrelativistic systems, so higher multipoles are inhibited.

where q is u or d. The lifetime of π^0 is $\sim 10^{-16}$ sec, a factor of $\sim 10^8$ shorter than that of $\pi^\pm$, and reflects the relative strength of the electromagnetic and weak interactions. Other decays of this type are $\eta \to 2\gamma$ and $\eta' \to 2\gamma$.

Examples of magnetic dipole ($M1$) transitions, satisfying the selection rules [Eqs. B(60) and B(61)], are $\gamma N \leftrightarrow \Delta$ ($\Delta J = 1$), $\Sigma^0 \to \Lambda^0 \gamma (\Delta J = 0)$, and $\psi' \to \eta_c \gamma$ ($\Delta J = 1$). In all these transitions, the only change of the multi-quark state is due to the spin flip of *one* quark, as one sees by comparing (6) with (8a) and (8b). The changing magnetic moment caused by this flip is responsible for these photoprocesses. Consequently, the $M1$ amplitudes are proportional to the quarks' intrinsic magnetic moments. In the case of the heavy c-quarks, the data is consistent with a magnetic moment μ_c having the Dirac value for a spin $\frac{1}{2}$ particle:

$$\mu_c = eQ_c/2m_c. \tag{12}$$

For light quarks the situation is not so clear-cut, because these systems are relativistic and (12) is no longer true. Nevertheless, one can attain a good understanding of the $M1$ transitions, and of the static magnetic moments of the lowest baryon levels N, Λ, Σ, and Ξ. As an example of the insight afforded by the model, consider the neutron-to-proton magnetic moment ratio, which is observed to be -0.685. If one assumes that u and d have intrinsic moments proportional to their charges, we may set

$$\mu_u = 2\mu_0/3 \quad \text{and} \quad \mu_d = -\mu_0/3. \tag{13}$$

According to Eq. (8a), p and n are $u_\uparrow u_\uparrow d_\downarrow$ and $d_\uparrow d_\uparrow u_\downarrow$, respectively. From this one readily shows* that $\mu_p = \mu_0$ and $\mu_n = -\frac{2}{3}\mu_0$, so that the ratio is $-\frac{2}{3}$, in excellent agreement with the facts.

According to (13) the intrinsic magnetic moments of the quarks μ_u and μ_d would then be $\frac{2}{3}\mu_p$ and $-\frac{1}{3}\mu_p$. This result is in reasonable agreement with an approximate relativistic generalization of (12), according to which the mass m_c should be replaced by the energy of the quark. Since three quarks make up a proton, the energy of one should be about $\frac{1}{3}$ of the proton mass m_p. Replacing m_c by this amount in (12) gives $\mu_q = 3Q\mu_B$, where $\mu_B = e/2m_p$ is the proton's Bohr magneton. From (13) we then get $\mu_0 = \mu_p = 3\mu_B$, which is not too far from the observed proton moment $2.8\mu_B$. Thus the magnetic moments of u and d are given by (13) with $\mu_0 \simeq \mu_p$.

---

* Let μ_{pair} be the moment of the parallel-spin pair; $\mu_{\text{pair}} = 4\mu_0/3$ and $-2\mu_0/3$ for $u_\uparrow u_\uparrow$ and $d_\downarrow d_\downarrow$, respectively. The pair has angular momentum 1, which combines with the spin of the third quark to give $J = \frac{1}{2}$. The values quoted for μ_p and μ_n then follow from the theorem that if a spin 1 system with magnetic moment μ_1 is combined with a spin $\frac{1}{2}$ system with magnetic moment $\mu_{\frac{1}{2}}$ to form a state of total angular momentum $\frac{1}{2}$, the system's magnetic moment is $(2\mu_1 - \mu_{\frac{1}{2}})/3$. Note that if the quarks were not in s-states, there would also be an orbital magnetic moment, whose contribution would spoil the relation $\mu_n/\mu_p = -\frac{2}{3}$. This is part of the evidence that the baryon ground state has $L = 0$, as mentioned on p. 86.

In order to compute the magnetic moments of other members of the baryon octet, we also need the magnetic moment μ_s of the s-quark. According to Eq. (8), the moment of the Λ is entirely due to s since the u–d pair has no angular momentum. Thus μ_s is the observed value $-0.61\mu_B$ of μ_Λ. Using these quark magnetic moments, one can compute the moments of the other members of the baryon octet. The comparison between theory and experiment is shown in Table 6.

TABLE 6

Baryon octet magnetic moments

Particle	Experiment	Quark Model
n/p	-0.685	-0.67
Σ^+	2.38 ± 0.02	2.7
Σ^-	-1.11 ± 0.03	-1.1
Ξ^-	-1.85 ± 0.75	-0.50
Ξ^0	-1.25 ± 0.01	-1.4
$\Sigma^0 \to \Lambda$	1.8 ± 0.2	1.6

All moments are in units of μ_B. In the $\Sigma^0 \to \Lambda$ case, the moment describes the $M1$ transition $\Sigma^0 \to \Lambda\gamma$. The Σ^+ and Σ^- moment measurements are new; Σ^+: Ankenbrandt (1983); Σ^-: Hertzog (1983).

Electron and muon scattering by hadrons is also a reaction involving the interaction of a photon with a hadron. The electron or muon produces a virtual photon γ which then interacts with the hadron [see Vol. II, Eq. III.A(22)]. Insofar as the hadron H is concerned, this causes a radiative transition of the type $\gamma H \to H'$, where H' is some other hadronic system which may consist of several hadrons. Elastic electron–hadron scattering is the important special case $H' = H$; as already mentioned, it can be used to determine the charge distribution of H.

(c) $e^+e^- \to$ hadrons

An important electromagnetic process that has been mentioned on several occasions is e^+e^- annihilation into hadrons. The first step, the annihilation of the e^+e^- pair into a virtual photon and the subsequent creation of a primary quark pair, is a purely electromagnetic process. The second step, the evolution of a number of color-neutral hadrons from the primary quark pair, is a strong interaction process. The total cross section of the reaction $e^+e^- \to$ hadrons is identical to the cross section $\sigma_{q\bar{q}}$ for producing primary pairs, since the second complex step follows the creation of a primary pair without interfering with its creation. Thus

$$\sigma(e^+e^- \to \text{hadrons}) = \sigma_{q\bar{q}} = \sum_{c,i} \sigma(e^+e^- \to q_{ci}\bar{q}_{ci}), \qquad (14)$$

where the sum runs over all those colors and flavors that can be created at the energy W of the e^+e^- pair. In order to include a flavor i in the sum of (14), W must be somewhat larger than the "flavor threshold" $2m_i \equiv W_i$, where m_i is the mass of q_i. For example, if W is between 4 and 10 GeV, the sum runs over $i = u, d, s, c$, but not b, since the flavor threshold W_c is ~4 GeV for c, and $W_b \sim 10$ GeV for b. The other flavor thresholds lie much lower.

The cross section (14) is readily determined. From deep inelastic scattering [recall §5(a)] we infer that quarks, like muons, behave as pointlike spin-$\frac{1}{2}$ particles at presently accessible momentum transfers. The cross section for production of a quark pair is then essentially the same as that for a muon pair,* $\sigma(e^+e^- \rightarrow \mu^+\mu^-)$. But there are differences.

The most important is the difference in charge. The amplitude for the transformation of γ into any pair is proportional to the charge of that pair [recall Eq. (10)], and therefore $\sigma(e\bar{e} \rightarrow q_{ci}\bar{q}_{ci}) = Q_i^2 \sigma(e\bar{e} \rightarrow \mu\bar{\mu})$, where $Q_i = \frac{2}{3}$ or $-\frac{1}{3}$. There is also the rather trivial difference in mass. We can ignore this by considering energies W that are above W_i by more than the quark mass of that flavor. Then the primary quarks are all relativistic, and their masses irrelevant. These considerations thus lead us from (14) to

$$R(W) \equiv \frac{\sigma(e\bar{e} \rightarrow \text{hadrons})}{\sigma(e\bar{e} \rightarrow \mu\bar{\mu})} = 3 \sum_i Q_i^2, \tag{15}$$

where the factor 3 counts the three colors, and the sum runs over the flavors whose thresholds W_i lie well below the energy W.

We therefore expect $R(W)$ to have a constant value, as given by (15), when W lies between two thresholds W_1 and W_2, but well above the lower one, W_1. Then it should rise to a higher plateau when W increases appreciably beyond W_2, so that the primary quarks are relativistic. As we shall see in §III.B.2 (Vol. II), this behavoir of R is observed experimentally.** The heights of the plateaus are in reasonably good agreement with the prediction of (15). This shows that a factor of about 3 is required in counting quarks of each flavor, and provides confirmation of the tri-valued color concept.

9. Charge-changing weak interactions

(a) General remarks

The weak interaction gives rise to a large set of phenomena in the subnuclear world. In nuclear physics its principal manifestation is the emission of an $e\bar{\nu}$ pair, and the simultaneous transformation of a neutron into a proton (β-decay). Other weak nuclear processes are related to this one by crossing. These phenomena involve a change of charge by both the hadrons (e.g., $n \rightarrow p$) and the leptons (vacuum $\rightarrow e\bar{\nu}$). In the subnuclear realm there

---

* This process is treated in greater detail in §II.C.2 (Vol. II).

** In the region near thresholds, new phenomena appear because of the bound states of the quark pairs, as discussed in §III.B.2 (Vol. II).

are many new counterparts to such charge-changing weak processes. But there are also weak processes in which neither the hadrons, nor the leptons, undergo a change of charge.* Therefore we divide the weak interactions into two categories: *charge-changing* and *charge-preserving* (or "neutral-current") weak interactions. The latter are discussed separately in the next section.

In §D we interpreted the emission of a lepton pair by a complex nucleus as being due to the underlying processes

$$n \to pe\bar{\nu} \quad \text{or} \quad p \to n\bar{e}\nu.$$

But we now know that nucleons are themselves complex structures, and so it is clear that these cannot be fundamental weak interaction processes. They must be regarded as manifestations of reactions involving the nucleon's constituents—of the transformation of a quark of one variety into one of another variety. Since the proton differs from the neutron by the replacement of one u by a d [see Eq. (8)], the β-processes can be interpreted as resulting from the following elementary transformations

$$d \to ue\bar{\nu}_e \text{ or } u \to d\bar{e}\nu_e. \tag{16}$$

We designate the neutrino which is emitted together with the positron by the symbol ν_e, since we soon will encounter neutrinos of different kinds.

The much larger set of weak interaction phenomena in the subnuclear realm has two sources: the existence of leptons other than e and ν_e; and the involvement of quarks other than u and d. Neither of these aspects shows up in the low-energy regime of nuclear physics because of the high masses of the other charged leptons and quarks.

(b) Other leptons

Atomic and nuclear physics deal with only two leptons—the electron e and its neutrino ν_e. The dominant decay of the charged pions reveals the existence of two further leptons, and their antiparticles:**

$$\pi^- \to \mu^- \bar{\nu}_\mu,$$
$$\pi^+ \to \mu^+ \nu_\mu. \tag{17}$$

The muon (μ) is a point-like spin-$\frac{1}{2}$ fermion which, to the best of our knowledge,[†] is a replica of the electron having a mass $207m_e$. It is not an excited state of the electron, because the decay $\mu \to e\gamma$ does not occur (branching ratio $\leq 10^{-10}$). The following two facts are established

---

* These "neutral current" weak interactions are not confined to subnuclear physics. They also produce effects in both atomic and nuclear physics, but these are more difficult to detect than those discussed in §10.

** The decays $\pi \to e\nu_e$, which also exist, are suppressed for reasons to be explained on p. 131.

[†] This evidence is discussed in detail in §§II.B and D (Vol. II).

empirically: (a) the neutrinos ν_μ emitted in $\pi \to \mu$ differ from those emitted in β-decay; (b) ν_μ is different from $\bar{\nu}_\mu$. That $\nu_\mu \neq \nu_e$ is shown by the observation that neutrinos from pion decay almost always produce muons, not electrons, when they interact with nucleons (see Fig. 29). Furthermore, ν's arising from π^+-decay produce μ^-, while ν's from π^--decay produce μ^+, which demonstrates that $\nu_\mu \neq \bar{\nu}_\mu$. It is also known that ν_μ has a gigantic mean free path in matter, corresponding to an interaction as weak as that of ν_e. Whereas the mass of ν_e is known to be less than $10^{-4}m_e$, the experimental upper limit of the ν_μ-mass is $\sim m_e$; it may even be zero. A detailed study of μ-decay [see Eq. (19) below] shows that the spin of ν_μ is $\frac{1}{2}$.

A still heavier charged lepton, called τ, has been discovered in the reaction

$$e^+e^- \to \tau^+\tau^-. \tag{18}$$

Its mass is $m_\tau = 1.78$ GeV, and it is point-like to a precision quite comparable to that of e and μ. It is accompanied by a neutrino, ν_τ, and it is known that $\nu_\tau \neq \nu_{\mu,e}$. At this time the upper limit on the mass of ν_τ is $\lesssim 250$ MeV.

The μ and the τ have a finite lifetime. While $\mu \to e\gamma$, and $\tau \to \mu\gamma$ or $\tau \to e\gamma$ are not observed, μ and τ do decay in virtue of the weak interaction. The μ has one dominant decay channel,

$$\mu^- \to e^- \bar{\nu}_e \nu_\mu; \tag{19}$$

its lifetime is 2×10^{-6} sec. The τ, being heavier, has a greater variety of decay channels, among the most prominent being

$$\tau^- \to \mu^- \bar{\nu}_\mu \nu_\tau, \tag{20a}$$

$$\to e^- \bar{\nu}_e \nu_\tau, \tag{20b}$$

$$\to \rho^- \nu_\tau. \tag{21}$$

Its lifetime is 5×10^{-13} sec. To each of the processes (19–21) there is, of course, an antiparticle decay, such as $\tau^+ \to \rho^+ \bar{\nu}_\tau$.

All these data show that we must enlarge the concept of lepton number N_l as defined in §D.4. There is not just one lepton number, but three: N_e, N_μ, and N_τ, with N_e being the quantity we previously called N_l. If $N(\nu_\mu)$ is the number of ν_μ's in a state, etc., then, for example,

$$N_\mu \equiv N(\mu^-) - N(\mu^+) + N(\nu_\mu) - N(\bar{\nu}_\mu). \tag{22}$$

To our present knowledge, *the three lepton numbers are separately conserved in all processes, whether weak, electromagnetic, or strong.* All the reactions (16) through (21) conform to this rule.*

---

* As an example of evidence for N_l conservation, consider the N_l-violating decay $K^0 \to \mu\bar{e}$. This has never been observed, the upper limit on the branching fraction being $\sim 10^{-6}$.

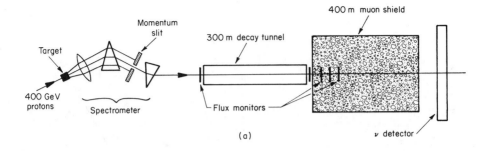

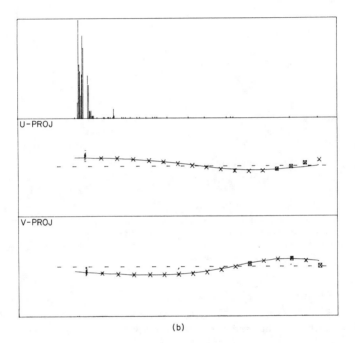

FIG. 29. (a) Sketch of the CERN neutrino beam (Dydak, 1978). A 400 GeV proton beam strikes a large-Z target, and produces a copious flux of pions, and some kaons. An array of magnetic lenses selects particles of a well-defined charge and mass, and within a specified spread of momentum. The hadron beam then passes through an empty region having a length such that $\pi \to \mu\nu$ decay is overwhelmingly probable, whereas μ decay is not (the lifetimes are related by $\tau_\mu \sim 10^2 \tau_\pi$). The beam then traverses several hundred meters of earth, which filters out the muons, but through which the neutrinos pass with impunity. Neutrino interactions are then observed in a complex device euphemistically called "detector" in the sketch. This can be a bubble chamber (see Fig. 33), or an electronic detector.

(b) A neutrino event as seen in the CERN-Dortmund-Heidelberg-Saclay detector (unpublished). This apparatus, which is immersed in a magnetic field, consists of a large sequence of detection planes perpendicular to the neutrino beam direction. These measure the ionization of charged particles passing through them, as well as the position in each plane where charged tracks have penetrated. The planes are separated by sheets of iron in which hadrons are slowed down and absorbed much more rapidly than muons. Consequently, the detector can single out the muon track. In these computer-drawn plots, the horizontal axis measures distance along the neutrino beam direction. The vertical axis in the uppermost plot shows the ionization deposited by hadrons as a function of depth into the detector. The two lower plots are orthogonal projections of the muon track; the spiraling orbit due to the magnetic field is clearly visible. This track goes beyond the end of the detector, whereas all hadrons are stopped well within the detector.

117

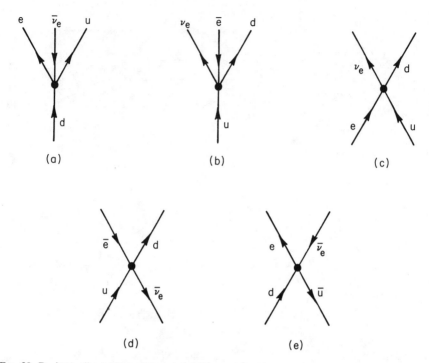

FIG. 30. Basic semileptonic processes involving quarks and leptons; see Eq. (23). β decay and positron emission by nuclei are due to the transitions (a) and (b), respectively, while K capture of atomic electrons by nuclei is due to (c). The reaction triggered by reactor neutrino beams is due to (d). The decay $\pi^- \to e\bar{\nu}_e$ is shown in (e).

It is an essential feature of weak interactions that the three lepton pairs play parallel roles. Processes that lead to the emission of one pair, also lead to the emission of the other pairs, if energy conservation permits.

(c) Role of other quarks

We now come to the second point which enriches subnuclear weak interaction phenomena, the involvement of other quarks. Their discussion will be considerably simplified by making use of the important crossing operation defined in §C.5. Crossing relates the two reactions (16), but it also implies the existence of many other processes related to β-decay:

$$d \to ue\bar{\nu}_e \quad u \to d\bar{e}\nu_e \quad eu \to d\nu_e$$
$$\bar{\nu}_e u \to \bar{e}d \quad d\bar{u} \to e\bar{\nu}_e \qquad \text{etc.} \tag{23}$$

Figure 30 illustrates these reactions in the form of diagrams.

Let us choose

$$d\bar{u} \leftrightarrow e\bar{\nu}_e \tag{24}$$

as the "representative" reaction from which all others in (23) are derived. In this reaction the particle pairs on either side have a total charge $Q = -1$, and zero baryon and lepton numbers: $B = N_e = 0$.

Among the fundamental fermions, there are many other particle–antiparticle combinations that have the quantum numbers carried by the pairs in (24), and therefore it is not surprising that there should exist many other weak reactions similar to (24). If l stands for *any* of the negatively charged leptons,

$$l = e^-, \mu^-, \text{ or } \tau^-,$$

and ν_l for the corresponding neutrino, while q_1 is any $Q = -\frac{1}{3}$ quark (d or s) and q_2 any $Q = \frac{2}{3}$ quark (u or c), all representative reactions of the type (24) are summarized by

$$q_1 \bar{q}_2 \leftrightarrow l \bar{\nu}_l \qquad (Q = -1, N_l = 0). \tag{25}$$

Reactions related thereto by crossing are $q_1 \rightarrow q_2 l \bar{\nu}_l$, $\nu_l q_1 \rightarrow l q_2$, $q_2 \rightarrow q_1 l \bar{\nu}_l$, etc.

Equation (25) and its crossing counterparts comprise all processes where a quark pair transforms into a lepton pair, or vice versa. In addition, there are also weak processes among the quarks themselves, and the leptons themselves. These representative reactions are

$$q_1 \bar{q}_2 \leftrightarrow q_3 \bar{q}_4 \qquad (Q = -1) \tag{26}$$

$$l \bar{\nu}_l \leftrightarrow l' \bar{\nu}_{l'} \qquad (Q = -1, N_l = N_{l'} = 0) \tag{27}$$

where q_3 and q_4 are quarks having the same charges as q_1 and q_2, respectively, but whose flavors may differ from those of q_1 and q_2. Thus (26) contains the reactions $d\bar{u} \leftrightarrow d\bar{c}$, $s\bar{c} \leftrightarrow d\bar{u}$, etc.*

Equations (25)–(27) state that *weak reactions exist between all combinations of quark and lepton pairs if their total charge $Q = -1$, and their lepton and baryon numbers vanish, and in all reactions derived therefrom by crossing.* Thus if f_α stands for any of the fundamental fermions—either a quark or a lepton—the most general representative weak interaction is

$$f_\alpha \bar{f}_\beta \leftrightarrow f_\gamma \bar{f}_\delta \qquad (Q = -1, B = N_l = 0). \tag{28}$$

As always, when we say that a reaction exists we mean that the transition matrix element is nonzero; naturally the process only occurs if energy and momentum conservation allow it.

---

* We largely ignore the flavor b, because these quarks do not play an imporant role in all but the largest energy transfers; the pairs $b\bar{u}$ and $b\bar{c}$ should, in principle, be included in (25) and (26).

(d) Weak interaction processes

The representative reactions (25)–(28) serve to distinguish three categories of weak processes:

(A) semileptonic: all reactions derived from (25)
(B) nonleptonic: all reactions derived from (26)
(C) leptonic: all reactions derived from (27)

We first consider the semileptonic category, in which a quark pair appears or disappears in the representative reaction. There are many semileptonic reactions, and it is therefore useful to subdivide them according to the changes of flavor (i.e., ΔS, etc.) produced by the creation of a $q_1\bar{q}_2$ pair (see Table 7). Since $q_1\bar{q}_2$ has $Q = -1$, we see that reactions where $\Delta S \neq 0$, $\Delta S = \Delta Q$ are allowed, whereas $\Delta S = -\Delta Q$ is forbidden. The data fully confirm this, as well as the analogous selection rule on ΔC.

TABLE 7

Flavor changes in semileptonic reactions

$q_1\bar{q}_2$	$d\bar{u}$	$d\bar{c}$	$s\bar{u}$	$s\bar{c}$
ΔS	0	0	-1	-1
ΔC	0	-1	0	-1
$|\Delta I_3|$	1	$\frac{1}{2}$	$\frac{1}{2}$	0

We shall give illustrative lists of semileptonic reactions, subdivided in accordance with the change of S and C; a far more complete list is to be found in Appendix I. In each case we show the underlying quark reaction, and indicate the corresponding diagram in Fig. 30 by a letter in parentheses.

(A) Semileptonic processes: $\Delta S = \Delta C = 0$

$$
\left.
\begin{array}{l}
n \to pe\bar{\nu}_e \qquad \Sigma^- \to \Lambda e\bar{\nu}_e \\
\pi^- \to \pi^0 e\bar{\nu}_e \qquad \Xi^- \to \Xi^0 e\bar{\nu}_e
\end{array}
\right\}
\quad
\begin{array}{ll}
d \to ue\bar{\nu}_e & \text{Fig. 30(a)} \\
\end{array}
$$

$$
\left.
\begin{array}{ll}
\pi^+ \to \pi^0 \bar{e}\nu_e & u \to d\bar{e}\nu_e & \text{(b)} \\
\nu_l n \to lp & \nu_l d \to lu & \text{(c)} \\
\bar{\nu}_l p \to \bar{l}n & \bar{\nu}_l u \to \bar{l}d & \text{(d)}
\end{array}
\right\}
\quad (29)
$$

In the decays $n \to p$, $\Sigma^- \to \Lambda$, and $\Xi^- \to \Xi^0$ the energy release is less than m_π. If that were not so, these decays would be dominated by strong pion emission.

Next we turn to semileptonic processes wherein either S or C changes:

(A) $\Delta S = \pm 1$, $\Delta C = 0$; or $\Delta S = 0$, $\Delta C = \pm 1$

$$\left.\begin{array}{l} \Lambda \to p l \bar{\nu}_l \\ K^- \to \pi^0 l \bar{\nu}_l \end{array}\right\} \qquad s \to u l \bar{\nu}_l$$

$$K^+ \to \bar{l} \nu_l \qquad\qquad u\bar{s} \to \bar{l} \nu_l \qquad\qquad (30)$$

$$\bar{\nu}_l p \to \bar{l} \Sigma^0 \qquad\qquad \bar{\nu}_l u \to \bar{l} s$$

$$\nu_l p \to l D^0 X^{++} \qquad\quad \nu_l d \to l c$$

The last reaction is the production of a charmed D meson together with a multihadron state X^{++} of $Q = 2$ in a ν–p collision; because of the stated selection rules, X^{++} has $I = \frac{1}{2}$ or $\frac{3}{2}$, and $C = S = 0$.

The energy differences between the initial and the final hadrons in the decays of Eq. (30) are larger than m_π, but strong pion emission is forbidden because $\Delta S \neq 0$, or $\Delta C \neq 0$. But these differences are less than the mass of the kaon, whose emission by the strong interaction would not be forbidden when $\Delta S = \pm 1$. Some of the reactions (30) are shown as diagrams in Fig. 31.

All lepton pairs emitted by strange baryons have $Q = -1$: thus $\Sigma^- \to n$ occurs, but $\Sigma^+ \to n$ cannot. This is so because s can only decay to u, which increases the hadronic charge, and the leptons must then carry off a compensating negative charge. Furthermore, there are no baryon decays with $\Delta S = -1$ or $\Delta C = +1$. They would require the transitions $u \to s$ or $d \to c$, which would increase the energy since $m_{s,c} > m_u$. But scattering processes with $\Delta S = -1$ or $\Delta C = +1$ are possible, such as the last two reactions in (30), because the energy is delivered by the incoming particle.

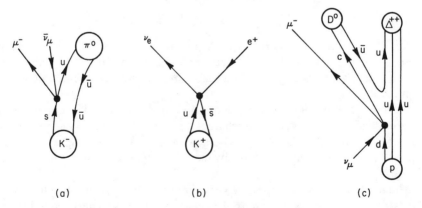

(a)   (b)   (c)

FIG. 31. Semileptonic reactions from among those listed in Eq. (30). A typical K^- decay is show in (a), and a K^+ decay in (b). The production of the charmed meson D^0 in a neutrino reaction is illustrated in (c).

The last category of semileptonic decays expected from Table 7 involves simultaneous changes of S and C:

$$\text{(A) } \Delta S = \Delta C = \pm 1$$

$$
\begin{aligned}
D^+ &\to K^- X^+ \bar{l} \nu_l & c &\to s \bar{l} \nu_l \\
F^+ &\to X^0 \bar{l} \nu_l & c\bar{s} &\to \bar{l} \nu_l,
\end{aligned}
\qquad (31)
$$

where X^+ and X^0 are multihadron states of the indicated charge, with $I = S = C = 0$.

Next we turn to the nonleptonic weak reactions (B) derived from Eq. (26). These fall into two categories: flavor conserving and flavor changing. An example of the former is $\bar{u}d \leftrightarrow \bar{u}d$ which, by crossing, implies the existence of a weak process $ud \leftrightarrow ud$. This will add a tiny weak contribution to the force between nucleons, and has actually been detected because it produces very small but measurable parity-violating effects in nuclear spectra and collisions.* However, such flavor-conserving weak effects are usually masked by the strong interaction. For that reason, we are principally interested in those nonleptonic reactions derived from (26) that violate strangeness or charm. These nonleptonic reactions (B) are remarkable because they allow pion emission in decays where such emission is forbidden to the strong interaction. Important examples are:

$$\text{(B) Nonleptonic processes:}$$

$$
\Lambda \to \begin{cases} n\pi^0 \\ p\pi^- \end{cases}
\qquad \Xi^- \to \Lambda\pi^- \quad \Omega^- \to \Lambda K^- \qquad s \to \bar{u}du
$$

$$
K^+ \to \pi^+ \pi^0 \qquad \bar{s} \to \bar{u}\bar{d}u
$$

$$
D^+ \to \bar{K}^0 \pi^+ \qquad c\bar{d} \to s\bar{d}, \; c \to su\bar{d}
\qquad (32)
$$

Several of these reactions are shown as Feynman diagrams in Fig. 32. In all these decays the mass differences between the hadrons are also smaller than m_K. If they were larger, kaon emission by the strong interaction would dominate.

In some nonleptonic decays (e.g., $\Lambda \to n\pi^0$) it would seem that the hadrons had not changed their charge, because the underlying process is $s \to u(d\bar{u})$. But this is also a charge-changing process, related to the representative reaction $s\bar{u} \to d\bar{u}$ by crossing, like all other processes in Eq. (32).

The nonleptonic decays (32) are very similar to the semileptonic decays. In the latter, lepton pairs are emitted instead of mesons. The probability for meson emission is 10^2–10^3 times greater than for lepton pair emission. This

* See, for example, Lockyer et al. (1980).

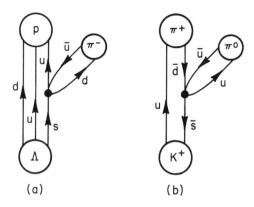

FIG. 32. Nonleptonic decays of a baryon (a) and a meson (b); see Eq. (32).

may raise doubts as to whether the nonleptonic decays (32) are really mediated by the weak interaction. However, there is overwhelming evidence that these are weak processes: they are parity violating, and do not conserve flavor. The much larger emission rate is not due to a fundamentally different interaction.*

Finally we come to the purely leptonic reactions (C), derived from (27). The most important are the μ- and τ-decays given in (19) and (20). Other reactions in this group are scattering processes between charged and uncharged leptons, such as

$$\nu_e e \rightarrow \nu_e e, \tag{33a}$$

$$\nu_\mu e \rightarrow \nu_e \mu. \tag{33b}$$

The first is elastic scattering between a neutrino and an electron, where the latter may be bound in an atom. Reaction (33b) is a neutrino-induced transformation of an electron into a muon. These elusive reactions can be detected by observing e and μ tracks in a bubble chamber exposed to an appropriate neutrino beam. As with all neutrino-induced reactions, the cross sections for (33) are very small. A rough formula for the cross section of all high energy ν reactions, whether on lepton or nucleon targets, is

$$\sigma_\nu \sim E \cdot 10^{-12} \text{ fm}^2, \tag{34}$$

where E is the laboratory neutrino energy in GeV. This area has a characteristic dimension that is far smaller than the radii of hadrons at all but

---

* The rate Γ_{if} for a transition $i \rightarrow f$ is given by $\Gamma_{if} = 2\pi|\langle i|H_{wk}|f\rangle|^2\rho_f$, where H_{wk} is the weak interaction, and ρ_f the density of final states. The latter depends strongly on the number of particles emitted, and is very different, for example, in $\Lambda \rightarrow p\pi^-$ as compared to $\Lambda \rightarrow pe^-\bar{\nu}_e$. The final multiquark states $|f\rangle$ also differ greatly. Detailed analysis shows that the terms in H_{wk} responsible for semileptonic and nonleptonic decays have the same magnitude.

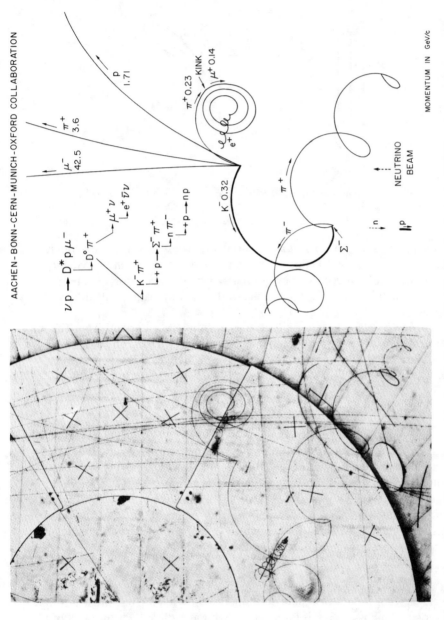

FIG. 33. A neutrino reaction, as observed in the Big European Bubble Chamber (courtesy CERN). This remarkable event displays virtually everything in the elementary particle lexicon: hadronic decays and collisions involving hadrons that carry charm, strangeness, or neither; semileptonic weak collision and decay processes; and leptonic as well as nonleptonic weak decays.

The photographed event is drawn on the right. The neutrino initially produces a charmed vector meson D^*. This decays hadronically to $D^0\pi^+$. The complete $\pi-\mu-e$ decay chain is shown. D^0 decays into a kaon, which collides with a proton in the chamber, producing Σ^-, which decays nonleptonically into $n\pi^-$. That neutron, whose track is invisible in the chamber, goes on to

124

extreme energies. Even at 10^3 GeV, it corresponds to a mean free path of 10^6 km in iron.

Figure 33 is an exceptional bubble chamber picture that displays all the phenomena we have discussed as a sequence of nonleptonic, semileptonic, and purely leptonic decays following a neutrino reaction, which is itself a semileptonic process.

(e) The intermediate vector bosons $W^\pm$

All the weak processes discussed thus far are derived by crossing from the representative reactions (28) involving the transformation of some fermion–antifermion pair $f_\alpha \bar{f}_\beta$, having $Q = -1$, into another pair $f_\gamma \bar{f}_\delta$ with the same charge. In these representative transmutations, the total baryon and lepton numbers of the participating pair vanish. The existence of this large set of reactions leads naturally to the hypothesis that *all those weak processes are manifestations of a more basic transformation wherein a fermion pair with $Q = \pm 1$ and $B = N_l = 0$ changes into a charged boson $W^\pm$, or vice versa*, as described by the following diagrams:

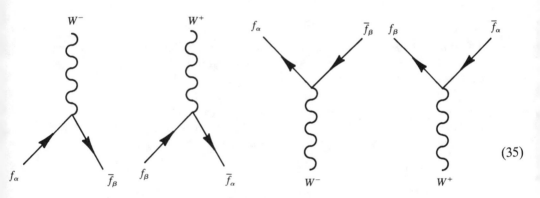

(35)

where W^+ and W^- are each other's antiparticles. By our hypothesis, they are *the quanta of a field that mediates the charge-changing weak interaction.*

Because of crossing symmetry, the pair creation and annihilation vertices also have $W^\pm$ emission and absorption counterparts:

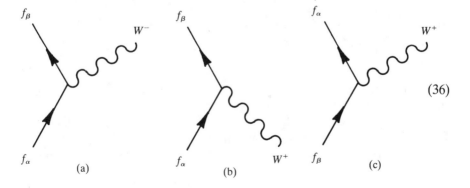

(36)

etc. By appropriately combining the one-vertex diagrams (35) and (36) one obtains all the 4-fermion processes discussed thus far:

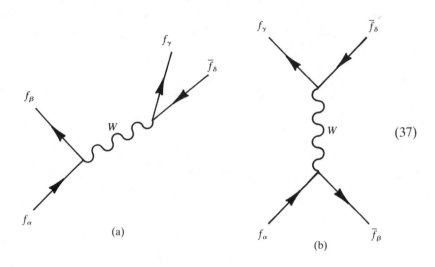

(37)

(a)   (b)

The dots in Figs. 30–32 are then replaced by wavy W lines. For example, the semileptonic decays, such as $K^- \to \pi^0 l \bar{\nu}_l$ shown in Fig. 31, are special cases of Eq. (37a), as are nonleptonic decays such as $K^+ \to \pi^+ \pi^0$, shown in Fig. 34(a). Decays involving quark pair annihilation such as $K^+ \to e^+ \nu_e$ shown in Fig. 31(b), are special cases of Eq. (37b). One-to-three fermion decays, such as $\mu^- \to \nu_\mu e^- \bar{\nu}_e$ or $\tau^- \to \rho^- \nu_\tau$, are also examples of (37a), while neutrino reactions, such as $\nu_\mu d \to \mu^- u$, are special cases of the diagram

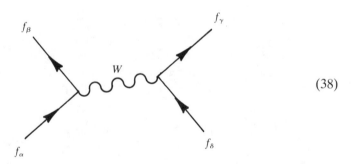

(38)

The charge of the intermediate W in these diagrams is always determined by the charges of the fermions, and is therefore not indicated.

These weak-interaction diagrams all have electromagnetic counterparts. For instance, the β-decay type diagram (37a) has the same structure as that for $e^+ e^-$ emission by nuclei, Eq. D(11). The annihilation graph (37b) has the same form as that for ψ-production in e^+e^- annihilation, as demonstrated in Fig. 34, which also shows how this mechanism works in $K^+ \rightarrow \pi^+\pi^0$.

Another remarkable similarity between the weak and electromagnetic fields is that both are vector fields. A detailed analysis of the angular distributions of weak processes indicates that W has spin one. A field whose quanta have spin one is a vector field, and for that reason W is called a vector boson.

Despite these intriguing analogies to electrodynamics, the differences between the weak and electromagnetic interaction are—at least at first sight—far more striking. Photons are readily produced, whereas W's can only be observed at the highest energies currently available [see §10(d)]. The electromagnetic interaction is long-ranged, causing scattering at large-impact parameters which leads to the familiar $(\sin\frac{1}{2}\theta)^{-4}$ singularity in the Rutherford formula for the scattering of charged particles [see §III.C.1 (Vol. II)]. In contrast, the angular distribution of weak collisions, such as the neutrino reactions described by (38), are essentially isotropic at all presently accessible energies. This indicates that the weak force has so short a range that collisions

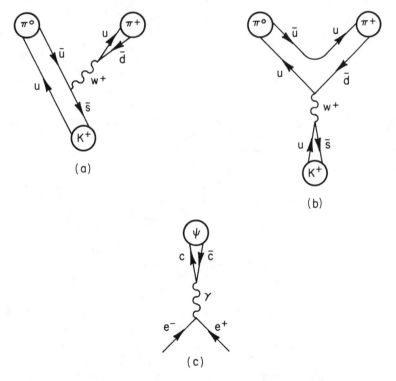

FIG. 34. Nonleptonic decays showing the role of the W boson. As shown, there can be more than one process contributing to a decay amplitude; in (a) the W is emitted in a fashion similar to bremsstrahlung, whereas in (b) it is created by pair annihilation. The latter is similar to meson production in $e\bar{e}$ annihilation, as shown in (c).

can only occur with very small impact parameters (or, equivalently, vanishing orbital angular momentum). An analysis of the angular distributions of decays of type (37a) also reveals that the distance between the point where the pair $f_\gamma \bar{f}_\delta$ is created, and the point where f_α transform into f_β, is immeasurably small.

These two contrasts are intimately related, because the range R of any force field is inversely proportional to the mass of the quanta of that field. If the weak force has an R that is small, m_W is large, and it takes energies of order m_W to produce $W^\pm$. The photon, being massless, has no threshold for its production. To understand why $R \sim 1/m_W$, consider the decay

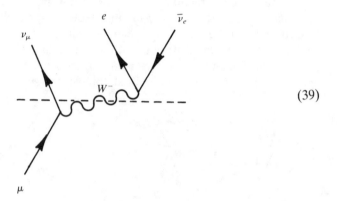

$$\tag{39}$$

The intermediate state, indicated by the dashed line, contains a W and neutrino. If W is far heavier than the muon, the energy difference ΔE between the initial and intermediate state is of order m_W. According to the uncertainty principle, the intermediate state exists for a time of order $1/\Delta E \sim m_W^{-1}$, during which interval a W cannot travel further than c/m_W. This demonstrates that *the distance R between the points where a W is emitted and reabsorbed is at most $\sim m_W^{-1}$ in all processes having a total energy small compared to m_W* (here we have reverted to $c = 1$).

Thus the two vertices in W-exchange graphs such as (37) and (38) are separated by a distance small compared to all other lengths (e.g., lepton or quark deBroglie-wavelengths) if the energies of the particles are low compared to m_W. Hence these processes appear to be due to an interaction where four fermions meet at a point, and this low-energy approximation is represented by the 4-fermion vertices in Figs. 30 to 32.

Recent discoveries, to be discussed in §10(d), provide a direct measurement of m_W, but it is actually possible to estimate its order of magnitude by exploiting the analogy between the weak and electromagnetic interactions. This analogy has been crafted into a theory that provides a unified description of weak and electromagnetic processes and that will be described in §10, and in greater detail in Vol. II, Chap. VI. For the moment a rough argument along the same lines will suffice. To this end, compare Rutherford scattering with a neutrino reaction:

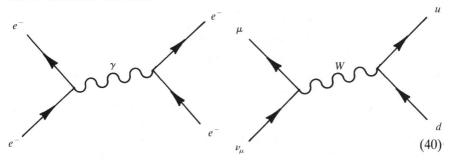

(40)

The amplitude whose square gives the Rutherford cross section is

$$A_R = \frac{e^2}{4\pi} \frac{1}{q^2},$$ (41)

where $q = 2E \sin \frac{1}{2}\theta$ is the momentum change suffered by an electron of energy $E(\gg m_e)$. The factor of e^2 arises because A_R is really a product of amplitudes, one for emitting the γ, the other for reabsorbing it, and these amplitudes are proportional to the electric charge of the emitter and absorber. If g is a "weak charge" carried by quarks and leptons that determines the amplitude for W absorption and emission, one finds that the neutrino reaction has the amplitude*

$$A_\nu = \frac{g^2}{4\pi} \frac{1}{m_W^2 + q^2} \simeq \frac{g^2}{4\pi m_W^2},$$ (42)

where q is the momentum difference between μ and ν_μ, which is always small compared to m_W at energies low compared to m_W. The measured magnitude (34) of neutrino cross sections shows that $g^2/m_W^2 \sim 10^{-4}$ GeV^{-2}. If we guess that the weak and electromagnetic charges g and e are of comparable magnitude, and that the low rates for weak processes are really due to the large mass of W, we find $m_W \sim 30$ GeV. In §10(d) we shall see that the presently measured value is

$$m_W \simeq 81 \text{ GeV},$$ (43)

which is in rough agreement with our crude estimate.

Hence W has a mass appreciably larger than that of the iron nucleus! The range of the weak interaction is therefore of order 10^{-3} fm, which is very

---

* Note that A_ν does not depend on angle if $E \ll m_W$, in agreement with our earlier argument. On the other land, for $E \gg m_W$, q^2 can also be much larger than m_W^2, and in this limit A_ν and A_R have the same form.

small compared to hadronic dimensions. For that reason this finite range cannot be observed in weak interaction phenomena having energies small compared to ~ 100 GeV.

(f) Space reflection and charge conjugation

Another remarkable difference between the weak and electromagnetic interaction is that the former violates parity. The manner in which this violation occurs can be succinctly stated by restricting oneself to kinematical situations where one or more fermions involved in a weak process are extremely relativistic ($v \simeq 1$). From a detailed analysis of angular distributions and correlations, and direct measurements of $e^{\pm}$ helicities, one has found the following rule: Whenever a charge-changing weak process leads to the creation of an extremely relativistic fermion f_{α}, or antifermion $\bar{f}_{\beta}$, their helicities are $h_{\alpha} = -\frac{1}{2}$ and $h_{\bar{\beta}} = \frac{1}{2}$, respectively. Since $i \to f$ implies $f \to i$, and because of crossing, this rule must apply to any fermion or antifermion, whether it is in the final state or not; in short, *the only helicity states of extremely relativistic fermions participating in any charge-changing weak process are $-\frac{1}{2}$ and $+\frac{1}{2}$ for particles and antiparticles, respectively.**

One might ask why this rule is restricted to extremely relativistic fermions. The reason is that *only for a massless particle is the helicity a Lorentz invariant.* This is illustrated by electrodynamics: a circularly polarized wave, corresponding to photons with $h = \pm 1$, has the same polarization in any Lorentz frame. For a particle with mass, a helicity eigenstate becomes a linear combination of various helicities when viewed from another frame.** As the description of any process must be frame-independent, it is only possible to have a firm rule concerning helicities of produced or destroyed particles if they are massless, or equivalently, in those kinematical domains where they move with $v \simeq c$, and their mass is irrelevant.

For nonrelativistic lepton momenta, the weak interaction is also parity violating. However, the magnitude of the violation is of order v/c in amplitude (see §VI.A.1, Vol. II).

The helicity rule just enunciated is, by its very nature, parity violating. Under a spatial reflection, $h \to -h$, and any phenomenon that prefers a helicity over its opposite is not reflection invariant (see §B.7). In the electromagnetic analogue of, say, $f_{\alpha}\bar{f}_{\beta} \to W \to f_{\gamma}\bar{f}_{\delta}$, such as $e^{+}e^{-} \to \gamma \to \mu^{+}\mu^{-}$, the lepton and antilepton helicities $\pm\frac{1}{2}$ interact with precisely the same strength. The weak interaction violates parity "maximally" because it only involves one of the fermion helicity states when $v = c$.

In §D.4 we already provided evidence for parity violation in β-decay, a semileptonic process. Parity violation is also observed in the other two weak

---

* Note the words "charge-changing" here, i.e., those processes described by W^{+} or W^{-} exchange. The rule does *not* apply to the so-called neutral current phenomena of §10.

** This can be seen simply by recognizing that a particle with a given helicity would have the opposite one when observed from a system that runs faster than the particle since there the momentum has changed sign, but not the spin direction.

reaction categories. Consider, for example, the nonleptonic decays $\Lambda \to N\pi$. The Λ's produced in certain parity-conserving strong reactions such as $\pi N \to \Lambda K$ have their spin partially polarized along a direction perpendicular to the reaction plane $\mathscr{P}$, as explained in Fig. 35. If parity is violated in $\Lambda \to N\pi$, the argument in §D.4 concerning the β-decay of Co^{60} shows that the pion's angular distribution will have an up–down asymmetry with respect to $\mathscr{P}$. Large asymmetries of this type are observed in this and other nonleptonic decays of baryons (e.g., $\Xi^- \to \Lambda\pi^-$). In the purely leptonic decay $\mu \to e\nu\nu$, there is a complete analogy to the evidence for parity violation in Co^{60} decay, because the electron angular distribution has an up–down asymmetry with respect to a plane normal to the muon spin.

The helicity rule accounts for the astonishing fact that $\pi^- \to e^- \bar{\nu}_e$ is a factor of 10^4 rarer than $\pi^- \to \mu^- \bar{\nu}_\mu$, even though the available energy is much larger in the first decay. In $\pi \to e$ decay, the electron's energy is 70 MeV $\simeq 140 \, m_e$, so this is an ultrarelativistic situation. In the pion's rest frame, e and $\bar{\nu}_e$ move in opposite directions. If their helicities were $-\frac{1}{2}$ and $\frac{1}{2}$,

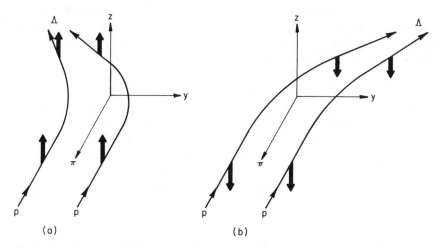

(a)   (b)

FIG. 35 Λ polarization in the parity-conserving reaction $\pi^-p \to \Lambda K^0$. For simplicity's sake, imagine that the only mechanism operative in the collision is of the (reflection-invariant) spin-orbit form. This force is taken to be attractive when the spin of the baryon is parallel to the relative orbital angular momentum between the baryon and meson, and repulsive when antiparallel. (Nothing is overlooked by ignoring spin-flip, because an unpolarized incident beam can always be viewed as 50% spin up and 50% spin down with respect to the reaction plane; then the only relevant term in the Hamiltonian is $\sigma_z L_z$, which cannot cause spin-flip.)

The collision is shown in the c.o.m. frame, with the incident pion and nucleon moving along the positive and negative x direction, respectively. In (a) the baryon spin, indicated by a thick arrow, is up. When the incident proton has a positive y intercept, $L_z > 0$, and there is an attraction, whereas $L_z < 0$ for a negative y intercept, giving a repulsion. In consequence, the classical (!) baryon trajectories are as shown in (a) for spin-up. The spin-down trajectories are in (b). With this simplistic interaction, all Λ's emerging on the left of the incident pion direction will have one spin direction, those on the right the opposite. That parity is not conserved in the subsequent weak decay can then be demonstrated by selecting a sample of Λ's produced in some angular region, and measuring the up–down asymmetry of decay pions with respect to the collision plane. Such an asymmetry requires parity violation by the argument of Fig. 4.

respectively, their spins would point in the same direction, and their combined angular momentum m along the decay direction would be $|m| = 1$, whereas $J_\pi = 0$! Hence $\pi \to e \bar{\nu}_e$ decay would be forbidden if e had the velocity c. In $\pi \to \mu$ decay, on the other hand, the μ moves slowly since $(m_\pi - m_\mu) \ll m_\mu$ (34 vs. 106 MeV). Such a slow μ^- is emitted in a state that is an almost equal mixture of "right" and "wrong" (i.e., $-\frac{1}{2}$ and $\frac{1}{2}$) helicities, and the "wrong" h combines with $h = \frac{1}{2}$ of $\bar{\nu}_\mu$ to give the required $J_\pi = 0$. In $\pi \to e$, the decay proceeds through the tiny portion (of order m_e/E) of the e^- state that has the "wrong" helicity.

Charge conjugation (C) invariance is also violated by the helicity rule. The action of C is to replace a particle by its antiparticle, while leaving its momentum and helicity untouched. If C (but not P) were a symmetry, the helicity rule would have to say that fermions *and* antifermions with the *same* helicity participate, whereas the data tell us that the participants are fermions and antifermions of the *opposite* helicity.

The decays $\mu^\pm \to e^\pm \nu\nu$ provide direct evidence for C violation. As already mentioned, there is an up–down asymmetry of the electron distribution in μ–e decay, demonstrating parity violation. If "up" is defined to be the μ spin direction, irrespective of whether it is a μ^+ or μ^-, C conservation requires the up–down asymmetry to be the same in $\mu^+ \to e^+$ and $\mu^- \to e^-$. It is not; instead the asymmetry changes sign.

The behavior of the weak interaction under space reflection and charge conjugation can also be stated very elegantly in terms of the spin and charge of the exchanged W's, as explained in Fig. 36. The figure also shows that even though C and P are both violated, *all the data summarized thus far are consistent with the statement that CP is a symmetry of the weak interaction*, or, put another way, the mirror image of a process looks precisely like that process in the anti-world.* As we shall learn in §11, this is not quite correct: CP is violated to one part per 10^3, but this violation is, at this time, only observable in K^0 decays.

(g) Universality

Thus far we have skirted the issue of the relative magnitude of various weak processes. Does the interaction responsible for $\mu \to e\nu\nu$ have the same strength as the one that causes β-decay? What is the relative size of strangeness-changing vs. strangeness-conserving decays? Are all the reactions described by (28) of the same strength?

We can orient ourselves by reexamining (10), the conversion of the electromagnetic quantum into quarks. Generalized to all the fundamental fermions, this reads

$$\gamma \to e \sum_\alpha Q_\alpha |f_\alpha \bar{f}_\alpha\rangle, \tag{44}$$

---

* An example of evidence for CP conservation is provided by the fact that the up–down asymmetry in $\mu \to e$ decay has the opposite sign to that in $\bar{\mu} \to \bar{e}$ decay, as explained in the preceding paragraph.

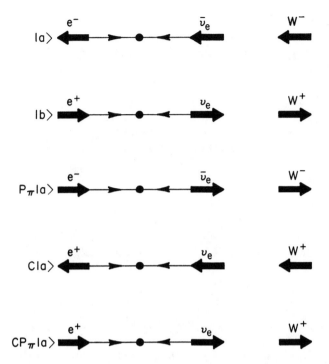

Fig. 36. The behavior of the weak interaction under P and C. All states illustrated contain a relativistic (e, ν_e) pair of zero total momentum. Thin and thick arrows are momenta and spins, respectively. The spin of the W's into which the leptons can annihilate is also indicated. The charge-changing weak interaction of the lepton pair (e, ν_e) only involves the states $|a\rangle$ and $|b\rangle$, or equivalently, $W^-(\leftarrow)$ and $W^+(\rightarrow)$. These states participate in the weak interaction with equal strength.

Let $\mathcal{S}$ be any symmetry operation. If $\mathcal{S}$ is a symmetry of the weak interaction, $\mathcal{S}|a\rangle$ must be either $|a\rangle$ or $|b\rangle$. Let P_π be a reflection through the origin, followed by a rotation of 180° about an axis perpendicular to the plane of the paper; this rotation is used so that the original momenta of $|a\rangle$ and $|b\rangle$ are recovered after reflection. $P_\pi|a\rangle$ is shown; it is neither $|a\rangle$ nor $|b\rangle$, and therefore the interaction is not reflection invariant (since rotational invariance is taken for granted). As the figure shows, C is also not a symmetry. On the other hand $CP_\pi|a\rangle = |b\rangle$, and also $CP_\pi|b\rangle = |a\rangle$. Consequently CP is a symmetry. However, as we shall learn in §11, there is a CP violation of order 10^{-3}, indicating that $|a\rangle$ and $|b\rangle$ have slightly different weak interactions.

where $Q_\alpha = \frac{2}{3}$, $-\frac{1}{3}$, or -1 are the charges of all known quarks and leptons. A similar expression can be written for the transformation of the weak-field quantum into fermions:*

$$W^- \rightarrow \frac{g}{\sqrt{2}} \sum_{\alpha\beta} a(\alpha,\beta)|f_\beta \bar{f}_\alpha\rangle. \tag{45}$$

* For reasons already explained in connection with Eq. (14), this formula, like (44), is only valid at energies large compared to the fermion masses. At lower energies other factors enter; they only depend on the fermion masses, and the total energy of the system [the quantity called W in §8(c)]. The factor of $\sqrt{2}$ is introduced for convenience.

Here g is the "weak charge" introduced in Eq. (42); it is the analogue of e in (44). The pure numbers $a(\alpha,\beta)$ give the relative size of the different amplitudes in W conversion; they are the analogues of the Q_α.

We already know that electromagnetism does not distinguish e from μ; the weak interaction has the same property, $a(\nu_e,e) = a(\nu_\mu,\mu)$, so by appropriately defining g one can set $a(\nu_e,e) = a(\nu_\mu,\mu) = 1$. This fact is established by comparing the decay rates for $A \rightarrow B\mu\bar{\nu}_\mu$ with those for $A \rightarrow Be\bar{\nu}_e$.* This symmetry of the weak interaction under the exchange $(e,\nu_e) \leftrightarrow (\mu,\nu_\mu)$ is called "$e\mu$ universality." Furthermore, lepton number conservation requires $a(\nu_e,\mu) = a(\nu_\mu,e) = 0$, and therefore

$$\begin{pmatrix} a(\nu_e,e) & a(\nu_e,\mu) \\ a(\nu_\mu,e) & a(\nu_\mu,\mu) \end{pmatrix} = \begin{pmatrix} 1 & 0 \\ 0 & 1 \end{pmatrix}. \tag{46}$$

Next we turn to the quarks. Their coefficients $a(\alpha,\beta)$ can also be written as an array like (46):

$$A = \begin{pmatrix} a(u,d) & a(u,s) \\ a(c,d) & a(c,s) \end{pmatrix}. \tag{47}$$

The data to be discussed show that one can redefine the quark states so that A also becomes the unit matrix (46). This is achieved by defining a new pair of orthogonal states for the $Q = -\frac{1}{3}$ quarks:**

$$|d'\rangle = |d\rangle \cos \theta_C + |s\rangle \sin \theta_C,$$
$$|s'\rangle = |s\rangle \cos \theta_C - |d\rangle \sin \theta_C, \tag{48}$$

where θ_C is called the Cabibbo angle. In this representation, the array becomes

$$\begin{pmatrix} a(u,d') & a(u,s') \\ a(c,d') & a(c,s') \end{pmatrix} = \begin{pmatrix} 1 & 0 \\ 0 & 1 \end{pmatrix}. \tag{49}$$

In short, the quark states u, d, c, and s are the degrees of freedom appropriate to the flavor-preserving strong interaction, whereas the states u, d', c, and s' are the degrees of freedom appropriate to the charge-changing weak interaction. That is to say, the quark states d', s' are the "normal modes" that diagonalize the bilinear form (45) for the weak transition

---

* Assuming $a(e,\nu_e) = a(\mu,\nu_\mu)$ one predicts the ratio of decay rates $(\pi \rightarrow e\nu):(\pi \rightarrow \mu\nu)$ to be 1.237×10^{-4}, whereas the measured value is $(1.267 \pm 0.023) \times 10^{-4}$. The observed lifetime of τ is also consistent with $a(\tau,\nu_\tau) = a(\mu,\nu_\mu)$, etc.
** The same result can be achieved by a corresponding transformation on the $Q = \frac{2}{3}$ quarks.

amplitude. As (46) and (49) demonstrate, *quarks and leptons have the same couplings to W's.* This is called quark-lepton universality.

Equations (48) and (49) allow us to rewrite (45) as follows:

$$W^- \to \frac{g}{\sqrt{2}}\left(|e\bar{\nu}_e\rangle + |\mu\bar{\nu}_\mu\rangle + |d'\bar{u}\rangle + |s'\bar{c}\rangle\right). \tag{50}$$

This expression displays the remarkably elegant and simple form of the basic charge-changing weak interaction.

Equations (47) and (48) imply that the array A is the orthogonal matrix

$$A = \begin{pmatrix} \cos\theta_C & \sin\theta_C \\ -\sin\theta_C & \cos\theta_C \end{pmatrix}. \tag{51}$$

Returning to quarks with definite flavors, the hadronic portion of (50) is therefore

$$W^- \to \frac{g}{\sqrt{2}}[\cos\theta_C(|d\bar{u}\rangle + |s\bar{c}\rangle) + \sin\theta_C(|s\bar{u}\rangle - |d\bar{c}\rangle)]. \tag{52}$$

Let us examine the data from which one infers that A becomes the unit matrix in the basis (48). Equation (52) says that for hadrons having no c-quarks, $\Delta S = \pm 1$ and $\Delta S = 0$ transition amplitudes are proportional to $\sin\theta_C$ and $\cos\theta_C$, respectively, and ratios of decay rates can therefore determine θ_C. Decays whose comparison are suited to this are $\Sigma^+ \to \Lambda e^+ \nu_e$ vs. $\Sigma^- \to n e^- \bar{\nu}_e$, and $K \to \mu\nu$ vs. $\pi \to \mu\nu$. A large number of such comparisons yield similar values of θ_C, and when combined lead to the result

$$\theta_C = 12.7° \pm 0.2°. \tag{53}$$

Because θ_C is small, $\Delta S = 1$ transitions are considerably weaker than $\Delta S = 0$ transitions for hadrons without c-quarks.

In itself, (53) does not establish quark–lepton universality, because θ_C was determined by the ratio $\tan\theta_C = a(u,s)/a(u,d)$. We also must show that $a(u,d') = 1$. This is done by measuring $|a(u,d)|^2$ and $|a(u,s)|^2$ separately. The former is found from the ratio of β-decay vs. μ-decay, and determines $\sin^2\theta_C$; the latter is fixed by the aforementioned measurement of $\tan\theta_C$. The result is

$$|a(u,d')|^2 = |a(u,d)|^2 + |a(u,s)|^2 = 0.997 \pm 0.032. \tag{54}$$

Finally we turn to decays involving c. Since θ_C is small, (52) predicts that in the decay of charmed hadrons, the dominant decays arise from

$W^- \to 2^{-\frac{1}{2}}g \cos \theta_C |s\bar{c}\rangle$, and therefore require $\Delta S = -1$, as in the semilepto-nic decays $c \to s\bar{l}\nu_l$. The data confirm this, for they show that the dominant electronic decay modes of D^+ are $D^+ \to K^- e^+ X^+$, where X^+ are $Q = 1$ multipion systems. The suppressed decays $c \to d\bar{l}\nu_l$, being proportional to $\sin \theta_C$, are difficult to measure, and there is as yet no direct proof, analogous to (54), that $a(c,s') = 1$.

⟦Nevertheless, there is indirect but highly compelling evidence that the second "charmed" row of A satisfies the condition required of an orthogonal matrix. Consider the almost forbidden decay $K^0 \to \mu^+\mu^-$, whose branching fraction is $\sim 10^{-8}$. This process can be caused by the exchange of two W's, as in

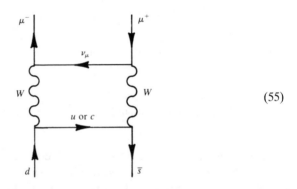

$\quad(55)$

Let A_u and A_c be the decay amplitudes when the quark u or c appears in the intermediate state. According to (45), each vertex in (55) has a factor $ga(\alpha,\beta)$, whence

$$A_u = g^4 a(u,s)a(u,d)f(m_u,m_W,\dots),$$
$$A_c = g^4 a(c,s)a(c,d)f(m_c,m_W,\dots),$$

$\quad(56)$

where ($\dots$) indicates a dependence on the masses of other particles involved in the diagrams. The complete decay amplitude is $A_u + A_c$, and as one see from (51), the orthogonal nature of A leads to

$$A_u + A_c = g^4 \cos \theta_C \sin \theta_C [f(m_u,m_W,\dots) - f(m_c,m_W,\dots)].$$

To the extent that m_u and m_c are negligible in comparison with m_W, the last factor vanishes, and the decay is forbidden. The cancellation is not quite complete, however, since $m_u \neq m_c$, and the small remainder gives a result consistent with the observed branching ratio for $K^0 \to \mu^+\mu^-$. On the other hand, if A were not orthogonal, the branching ratio would be orders of magnitude larger.⟧

10. The relationship between weak and electromagnetic interactions; neutral current weak processes

The theory of charge-changing weak interactions, based on the exchange of $W^\pm$ bosons between fermions, provides an elegant, economical, and highly successful description of a vast body of data, as indicated by the foregoing review. Nevertheless, this theory cannot be correct as it stands. First, weak processes have been discovered that can *not* be due to the basic mechanism* (45) wherein a *charged* W transforms into a pair $f_\alpha \bar{f}_\beta$ having $Q = \pm 1$, but which require a transition between a *neutral* boson and a fermion pair $f_\alpha \bar{f}_\alpha$ of $Q = 0$. Second, a theory with only two oppositely charged intermediate bosons leads to grave inconsistencies. It is not possible to discuss these problems with the limited amount of quantum field theory that we have developed so far, but they will be considered in §VI.A.3 (Vol. II).

(a) Weak isospin

An analysis of these undesirable features (see §§IV.A and B, Vol. II) reveals that they stem from a lack of symmetry of the charge-changing weak interaction. These symmetry features of weak processes can be understood by exploiting yet again the observation, made in §B.3(e), that any two-state system is mathematically equivalent to a spin, and that the unitary transformations in that two-state space $\mathscr{C}_2$ are equivalent to rotations in a Euclidean 3-space $\mathscr{E}_3$.

In the problem at issue the two-state system is any one of the fermion doublets (ν_e, e), (ν_μ, μ), (u, d') or (c, s') created by W^- or W^+ [recall Eq. (50)]. As these doublets appear on a completely symmetric footing we can, without loss of generality, focus our attention on the (ν_e, e) doublet. We then introduce a *weak isospin*, with eigenvalues $T = 0, \frac{1}{2}, 1, \ldots$, and $T_3 = T$, $T - 1, \ldots, -T$, and assign $T = \frac{1}{2}$ to this doublet, with $T_3 = \frac{1}{2}$ and $-\frac{1}{2}$ for $|\nu_e\rangle$ and $|e\rangle$, respectively. Weak isospin is mathematically equivalent to ordinary spin, and to the hadronic isospin carried by the (u, d) quark doublet.** Nevertheless, hadronic and weak isospin are *distinct* attributes, since leptons have nothing to do with hadronic isospin. The antidoublet $(\bar{e}, \bar{\nu}_e)$ is obtained from the doublet by reversing the charge, lepton number, and T_3; that is, the antidoublet also has $T = \frac{1}{2}$, with $\bar{e}$ and $\bar{\nu}_e$ having the eigenvalues $T_3 = \frac{1}{2}$ and $-\frac{1}{2}$, respectively (recall here the analogous situation for $\bar{d}$ and $\bar{u}$, as shown in Fig. 10). The *weak isospin space* will be designated by $\mathscr{C}_3^T$.

---

* As always, we speak here of the "representative" reactions as defined in §9(c); for an explicit statement in the present context, see the discussion following Eq. (60).

** Of the quark doublets created by the annihilation of W, one resembles the hadronic isospin doublet (u, d). This is the weak isospin doublet $(u, d')_L$, where d' is given by Eq. (48), and the subscript L indicates that when the motion is ultrarelativistic, *only* the $h = -\frac{1}{2}$ states of the quarks (and $h = \frac{1}{2}$ states of the antiquarks) contribute to the weak isodoublet, whereas *all* helicity states contribute to the hadronic isospin doublet. A detailed discussion of this distinction will be found in §VI.B.1(b), Vol. II.

The weak interaction involves fermion–antifermion pairs. For our purpose it is necessary to represent such pairs by eigenstates $|\Phi_T^{T_3}\rangle$ of the total weak isospin of the pair. This is readily done by copying from Table 4 for mesons built from (u,d) quarks; as always there is a triplet $|\Phi_1^{T_3}\rangle$ and a singlet $|\Phi_0\rangle$:

$$\left.\begin{aligned}
|\Phi_1^1\rangle &= |\nu_e\bar{e}\rangle, \\
|\Phi_1^0\rangle &= \frac{1}{\sqrt{2}}(|\nu_e\bar{\nu}_e\rangle - |e\bar{e}\rangle), \\
|\Phi_1^{-1}\rangle &= |e\bar{\nu}_e\rangle,
\end{aligned}\right\} \tag{57}$$

$$|\Phi_0\rangle = \frac{1}{\sqrt{2}}(|\nu_e\bar{\nu}_e\rangle + |e\bar{e}\rangle). \tag{58}$$

Equation (50), and its counterpart W^+, read as follows in this language

$$|W^\pm\rangle \rightarrow \frac{g}{\sqrt{2}}|\Phi_1^{\pm 1}\rangle. \tag{59}$$

As we see, the charge-changing weak interaction only involves *two* members of the $T = 1$ triplet. The weak interaction can only be invariant if the third member, $|\Phi_1^0\rangle$, is involved in the interaction with equal strength. As we see from (57), $|\Phi_1^0\rangle$ is electrically neutral ($Q = 0$), and it can therefore be produced by the annihilation of an electrically neutral field quantum W^0. In short, if the weak interaction is to be invariant under weak isospin rotations, we must introduce a third field to mediate the weak interaction, and complement (59) by a *new basic transition from the neutral boson W^0 to a neutral fermion–antifermion pair*:

$$|W^0\rangle \rightarrow \frac{g}{\sqrt{2}}|\Phi_1^0\rangle. \tag{60}$$

As always, the transition in (60) is a "representative" reaction, in the sense defined after Eq. (25). Thus (60) also describes processes such as $e^- \rightarrow W^0 e^-$, $W^0 \bar{\nu}_e \rightarrow \bar{\nu}_e$, as well as $\nu_e \bar{\nu}_e \rightarrow W^0$, etc. Furthermore, one should not forget that there are other weak isospin doublets, such as (ν_μ,μ), (u,d'), etc., allowing processes such as $\nu_\mu \rightarrow W^0 \nu_\mu$, $uW^0 \rightarrow u$, etc. As we see, the fermion charge does not change in these transitions, as it does in the processes involving $W^\pm$. For that reason we may call reactions involving W^0 *charge-preserving weak processes*.

(b) The electroweak connection and the charge-preserving weak interaction

Let us now turn to electromagnetism. It is obvious that this interaction does not conserve weak isospin, for it only involves e, the $T_3 = -\frac{1}{2}$ member of the (ν_e, e) doublet.* Consequently our guiding principle—invariance in the weak isospin space—does *not* allow us to add the electromagnetic interaction to the symmetric weak interaction involving W^+, W^0, and W^-. On the other hand, there is one linearly independent fermion–antifermion state which has not yet been coupled to any field quantum, the *invariant* $(T = 0)$ combination $|\Phi_0\rangle$. Because it is a scalar under weak isospin rotations, the interaction of this electrically neutral combination will not spoil what we have already done. We therefore introduce a new field, with an electrically neutral quantum B^0, which, on annihilation, produces $|\Phi_0\rangle$:

$$|B^0\rangle \to \sqrt{2}g'\langle Q\rangle|\Phi_0\rangle. \tag{61}$$

Here g' is a coupling strength analogous to e or g, the factor $\sqrt{2}$ is an arbitrary but convenient convention, and $\langle Q\rangle$ is the mean charge of the doublet in question, i.e., $\langle Q\rangle = -\frac{1}{2}$ for leptons and $\frac{1}{6}$ for quarks. Note that (61) does not distinguish between the members of a T doublet.

In sum, we now have four basic transmutations between fields and fermion pairs, as given by Eqs. (59–61), and full symmetry. That symmetry implies that W^0 and B^0 are spin 1 quanta, like W^+ and W^-.

The basic electromagnetic transmutation is given by (44) in all generality. For a lepton doublet this can be abbreviated to

$$|\gamma\rangle \to -e|e\bar{e}\rangle. \tag{62}$$

But $|\Phi_0\rangle$ and $|\Phi_1^0\rangle$ are linear combinations of $|e\bar{e}\rangle$ and $|\nu_e\bar{\nu}_e\rangle$, and therefore (60) and (61), with $\langle Q\rangle = -\frac{1}{2}$, can be combined to form the $|e\bar{e}\rangle$ state produced in the electromagnetic interaction:

$$\frac{g|B^0\rangle + g'|W^0\rangle}{\sqrt{g^2 + g'^2}} \to \frac{1}{\sqrt{2}}\frac{gg'}{\sqrt{g^2 + g'^2}}[|\Phi_1^0\rangle - |\Phi_0\rangle]$$

$$= -\frac{gg'}{\sqrt{g^2 + g'^2}}|e\bar{e}\rangle. \tag{63}$$

Here the factor $(g^2 + g'^2)^{-\frac{1}{2}}$ normalizes the state to unity. On comparing to (62) we have

$$|\gamma\rangle = \frac{g|B^0\rangle + g'|W^0\rangle}{\sqrt{g^2 + g'^2}}. \tag{64}$$

Equation (64) says that the photon is a coherent superposition of two states, one of which is on a symmetric footing with the charged states $W^\pm$

---

* In the case of quark doublets both members participate in electrodynamics, but asymmetrically, because they have charges $\frac{2}{3}$ and $-\frac{1}{3}$; see below.

responsible for the charge-changing weak processes.* Equations (62) and (63) also imply that the charge on the electron, the "weak charge" unit g, and the constant g' introduced in (61) are related by

$$e = \frac{gg'}{\sqrt{g^2 + g'^2}}. \tag{65}$$

This is a universal relation between e, g, and g'.

[In the case of a quark doublet $[(u,d')$, for example] the derivation proceeds as follows. As in (57) and (58), we define $T_3 = 0$ states with $T = 1$ and $T = 0$:

$$|\Phi_1^0\rangle = \frac{1}{\sqrt{2}} (|u\bar{u}\rangle - |d'\bar{d}'\rangle), \tag{57'}$$

$$|\Phi_0\rangle = \frac{1}{\sqrt{2}} (|u\bar{u}\rangle + |d'\bar{d}'\rangle). \tag{58'}$$

But (62) now reads

$$|\gamma\rangle \rightarrow e[\tfrac{2}{3}|u\bar{u}\rangle - \tfrac{1}{3}|d'\bar{d}'\rangle]. \tag{62'}$$

Combining (60) and (64) with (61), where $\langle Q \rangle = \tfrac{1}{6}$, we get instead of (63):

$$|\gamma\rangle = \frac{g|B^0\rangle + g'|W^0\rangle}{\sqrt{g^2 + g'^2}} \rightarrow \frac{gg'}{\sqrt{g^2 + g'^2}} (\tfrac{2}{3}|u\bar{u}\rangle - \tfrac{1}{3}|d'\bar{d}'\rangle) \tag{63'}$$

which, indeed, is the correct coupling of quarks to photons by virtue of (65).]

We have introduced two new neutral field quanta, B^0 and W^0, and shown that one linear combination, Eq. (64), is the photon. Consider the orthogonal linear combination, called Z^0:

$$|Z^0\rangle = \frac{g|W^0\rangle - g'|B^0\rangle}{\sqrt{g^2 + g'^2}}. \tag{66}$$

Since $|Z^0\rangle$ is linearly independent of the photon, there are two distinct interactions wherein field quanta transform into electrically neutral fermion pairs. One is the electromagnetic interaction, as represented by (62) or by (63'); the other is *a weak charge-preserving interaction in which the quantum Z^0 annihilates into fermion pairs.*

---

* As the admonition in the footnote related to Eq. (45) emphasizes, these relations between quanta are only correct at energies large compared to all masses. In §VI.B.2, Vol. II we shall show that (64), as it stands, is a generally valid relation between fields [see Eq. VI.B(47)].

By using Eqs. (60), (61), and (66) we may construct the fermion state that is created by this Z^0 annihilation:

$$|Z^0\rangle \rightarrow \frac{1}{\sqrt{g^2 + g'^2}} \cdot \frac{1}{\sqrt{2}} [g^2 |\Phi_1^0\rangle - 2g'^2 \langle Q \rangle |\Phi_0\rangle]. \tag{67}$$

It is instructive to write this as a linear combination of $|\Phi_1^0\rangle$, and the state $|\Phi_{em}\rangle$ produced by γ annihilation. For any given doublet (f_1, f_2), the latter can be written in the notation of (44):

$$|\Phi_{em}\rangle = \sum_{\alpha=1,2} Q_\alpha |f_\alpha \bar{f}_\alpha\rangle. \tag{68a}$$

In this notation the state $|\Phi_1^0\rangle$ is

$$|\Phi_1^0\rangle = \sqrt{2} \sum_{\alpha=1}^{2} T_{3\alpha} |f_\alpha \bar{f}_\alpha\rangle, \tag{68b}$$

where $T_{3\alpha} = \frac{1}{2}$ and $-\frac{1}{2}$ for $\alpha = 1$ and 2 respectively. The desired result can be obtained by adding and subtracting $g'^2 \langle Q \rangle |\Phi_1^0\rangle$ inside the brackets [ ] of Eq. (67); on using (57) and (58) one finds

$$|Z^0\rangle \rightarrow \sqrt{\tfrac{1}{2}(g^2 + g'^2)} |\Phi_1^0\rangle - \frac{g'^2}{\sqrt{g^2 + g'^2}} |\Phi_{em}\rangle.$$

It is customary to express this in terms of the electroweak mixing angle θ_W (also referred to as the Weinberg angle), defined by

$$\sin \theta_W = \frac{g'}{\sqrt{g^2 + g'^2}}. \tag{69}$$

In terms of θ_W, $|Z^0\rangle$ is*

$$|Z^0\rangle \rightarrow \frac{g}{\cos \theta_W} \left[\frac{1}{\sqrt{2}} |\Phi_1^0\rangle - \sin^2 \theta_W |\Phi_{em}\rangle \right].$$

This Z^0 transmutation law applies to all doublets. It can be written in a more convenient form by using the expressions for the fermion states given

---

* In this derivation, we have purposely glossed over the fundamental distinction between the weak and electromagnetic interactions insofar as parity violation is concerned. In particular, the states $|\Phi_T^{T_3}\rangle$ only contain fermions of $h = -\frac{1}{2}$, and antifermions of $h = \frac{1}{2}$, and are therefore not states of well-defined parity, whereas $|\Phi_{em}\rangle$ has the odd parity of the photon. Despite this "oversight," the final result (70) is correct. A complete derivation is given in §VI.B.3 (Vol. II).

in (68):

$$|Z^0\rangle \rightarrow \frac{g}{\cos \theta_W} \sum_\alpha (T_{3\alpha} - Q_\alpha \sin^2\theta_W)|f_\alpha \bar{f}_\alpha\rangle. \tag{70}$$

Hence any member of a weak isodoublet has a coupling to Z^0 that is proportional to its eigenvalue of $(T_3 - Q \sin^2\theta_W)$.

The Z^0 transmutation law (70) explicitly displays the *electroweak connection: The charge-preserving weak interaction is completely fixed by the electromagnetic interaction, and the charge-changing weak interaction.* In principle, the latter determines the coupling constant g, and also the state $|\Phi_1^0\rangle$, for it is a member of the same multiplet as the states $|\Phi_1^{\pm 1}\rangle$ involved in $W^\pm$ processes. The electromagnetic interaction then determines θ_W, because (65) and (69) imply $e = g \sin \theta_W$, while the state $|\Phi_{em}\rangle$ is found from the annihilation of γ. Nothing else enters into the amplitudes for Z^0 creation, or annihilation, as (70) shows. The situation is not so simple at energies well below 100 GeV, however. In that regime all charge-changing reaction amplitudes are proportional to g/m_W, and g cannot be measured directly (see §VI.A, Vol. II). Nevertheless, as we shall see shortly, tests of the electroweak theory are still possible at such energies.

(c) Symmetry breaking

Symmetry in the weak isospin space $\mathscr{E}_3^T$ leads inexorably to an electroweak connection. On the basis of that symmetry one would expect the members of the weak isotriplet (W^+, W^0, W^-) to be degenerate in mass, and the singlet B^0 to have some other mass. Yet one linear combination of W^0 and B^0, the photon, is massless; the orthogonal combination Z^0, and $W^\pm$, have masses of order 100 GeV. In short, the mass spectrum of the field quanta does not display the assumed symmetry of the electroweak interaction.

This puzzling situation occurs in many physical systems, for example, in ferromagnetism. The Hamiltonian for an iron crystal is invariant under spatial rotations. Nevertheless, the ground state is not invariant: It distinguishes a specific direction—the direction of magnetization. To an observer living inside the iron crystal, the rotational symmetry would be hidden; only appropriate measurements, combined with a sound understanding of ferromagnetism, would allow such an observer to discover the rotational invariance of the dynamical laws that govern his environment. When the temperature (i.e., the energy) of the crystal is raised above the Curie point, the magnetization disappears, and the rotational symmetry becomes manifest. One says that the symmetry is spontaneously broken in the low-temperature phase, a term borrowed from the term "spontaneous magnetization." As we have seen, the symmetry actually survives, but it is hidden from view.

In the current theory, the weak isospin symmetry breaking is introduced by the so-called Higgs mechanism. One postulates the existence of a Higgs field, which is a scalar under spatial rotations (i.e., has spin zero quanta), but is a weak isodoublet. This field is coupled in a *symmetric* manner to quarks, leptons, W's, and Z's. *The essential assumption is that the ground state of the system—the vacuum—is asymmetric, even though the basic Hamiltonian of the system is fully symmetric.* This vacuum asymmetry emerges from a Hamiltonian that is $\mathscr{E}_3^T$-symmetric, but having a minimum when a member of the Higgs doublet (say the one with $T_3 = -\frac{1}{2}$) has a *nonzero* value in the vacuum. A particular direction in the space $\mathscr{E}_3^T$ is therefore singled out, and this hides the weak isospin symmetry. The analogy to ferromagnetism is self-evident: The Higgs field with a nonvanishing vacuum expectation value plays the role of the magnetic field.

This picture adds a new feature to the vacuum. All other fields—the electromagnetic, weak, and strong fields, and those that describe the fermions—have fluctuations about a vanishing expectation value in empty space.

Because of the vacuum asymmetry, the phenomena that stem from the coupling of the Higgs field to the fermions and bosons no longer manifest the underlying, hidden, weak isospin symmetry. At energies large compared to 100 GeV, all masses are negligible and the weak isospin symmetry is manifest, but at low energies it is hidden. This is especially true of the masses of the field quanta, because the coupling of the Higgs field to the vector fields that mediate the electroweak interaction is arranged so as to give the W and Z^0 masses in the 100 GeV range, while maintaining the photon mass at zero. Despite this apparent symmetry breaking, low-energy experiments on weak and electromagnetic phenomena can reveal the underlying weak isospin symmetry, just as appropriate observations inside a ferromagnet can uncover the hidden rotational symmetry.

Whether the Higgs mechanism is actually the correct explanation of the breaking of weak isospin symmetry that we observe is not at all clear at this time. As it stands, this theory predicts the existence of an electrically neutral spin-zero "Higgs particle," but it cannot predict its mass. No particle answering to this description has been found as yet.

The theory of symmetry breaking that we have just sketched (the so-called Weinberg–Salam model) predicts a definite ratio between the masses of W and Z^0 determined by the electroweak mixing angle:

$$\frac{m_W}{m_Z} = \cos \theta_W. \tag{71}$$

Neither the coupling constants g and g', nor the actual values of m_W and m_Z, are determined by the theory. What the theory does predict is that the mass ratio (71), and the coupling constant ratio e/g [recall Eq. (69)], satisfy

$$\left(\frac{m_W}{m_Z}\right)^2 + \left(\frac{e}{g}\right)^2 = 1. \tag{72}$$

As we have already emphasized, the theory also specifies the complete structure of the Z^0-mediated weak interaction in terms of the electromagnetic and $W^{\pm}$-mediated weak interactions.

(d) Determination of θ_W, m_W, and m_Z. Observation of W and Z

Finally we come to the determination of the masses of the intermediate bosons, and of the angle θ_W. At energies small compared to M_W, the amplitudes for charge-changing weak processes are proportional to $(g/m_W)^2$, as we saw in Eq. (42). Hence the measurements of weak decay rates (e.g., for $\mu \rightarrow e\nu\nu$) determine this ratio, and tell us that $(m_W/g) = 123$ GeV. By using $g = e/\sin \theta_W$, and $4\pi/e^2 = 137$, one obtains $m_W = (37.3/\sin \theta_W)$ GeV. At the level of accuracy required by current measurements, corections due to higher order perturbations arising from the weak and electromagnetic interactions must be included. These have been computed [see Kim (1981); Marciano (1983)] and give the final result

$$m_W = \frac{38.7}{\sin \theta_W} \text{ GeV.} \tag{73}$$

The parameter θ_W can be determined from Z^0-mediated weak processes. The basic Z^0 fermion pair vertices are

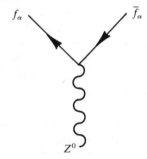

where f_α is any lepton or quark. As is clear from (70), these amplitudes for $Z^0 \leftrightarrow f_\alpha \bar{f}_\alpha$ depend on θ_W, and also on the weak isospin and charge of the fermion f_α. By combining two vertices of this type we obtain a host of charge-preserving "neutral-current" weak processes. Among these, consider in particular

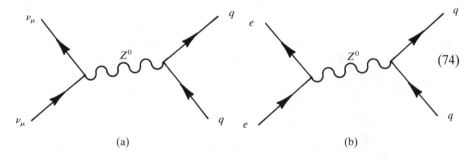

$$\tag{74}$$

(a)   (b)

where q is any quark. Here (a) is a neutrino collision in which there is a neutrino in the final state, in contrast to the "charged current" reactions mediated by $W^{\pm}$ exchange with a charged lepton in the final state. The mechanism 74(a) gives rise to so-called "neutral-current" neutrino reactions $\nu A \to \nu X$, where A is a nucleus, and X some hadronic state.* Diagram 74(b) gives a contribution to electron-nucleon scattering; in contrast to the dominant electromagnetic process due to γ-exchange, it is parity violating, and can therefore be detected by measuring the difference in cross sections for electrons with helicity $\frac{1}{2}$ and $-\frac{1}{2}$.

At this time the most accurate determination of θ_W comes from measuring the following ratios of total cross sections:

$$\frac{\sigma(\nu_\mu A \to \nu_\mu)}{\sigma(\nu_\mu A \to \mu)} \quad \text{and} \quad \frac{\sigma(\bar{\nu}_\mu A \to \bar{\nu}_\mu)}{\sigma(\bar{\nu}_\mu A \to \bar{\mu})},$$

where A is a complex nucleus. These ratios are known functions of the parameter $\sin^2\theta_W$, as we shall demonstrate in §VI.B.3 (Vol. II). In this way one finds that**

$$\sin^2\theta_W = 0.217 \pm 0.014 \tag{75}$$

The rather high accuracy of this result is due to the very large number of neutral current neutrino events that have been observed. An independent determination of $\sin^2\theta_W$ is provided by the observed magnitude of the parity violating asymmetry in eN scattering; this gives $\sin^2\theta_W = 0.223 \pm 0.015$, in agreement with (75). There are also several other independent measurements of $\sin^2\theta_W$; they are of lower accuracy, but consistent with the two results that we have quoted.[†]

These experiments also allow one to extract the ratio $m_W/m_Z \cos \theta_W$, even though they do not determine the masses (see §VI.B.3, Vol. II). The result is

$$\frac{m_W}{m_Z \cos \theta_W} = 1.02 \pm 0.02$$

in excellent agreement with (71).

---

* A "neutral-current" neutrino event is differentiated from events such as the ones shown in Figs. 29(b) and 33 in that there is no muon track, only a hadronic shower. We should also remark that the word "current" is apt in these weak processes, for reasons to be explained in Vol. II, Chap. VI. On the other hand, the terminology "charged" and "neutral" has no rationale, as we shall see there; like many other bits of unfortunate jargon, they seem to have gained a permanent niche.

** The results quoted here are from the review of Marciano (1983), where references to the original measurements can be found. Also see Kim (1981).

† The neutrino and eN data, when taken together, also confirm that the coupling of the various quarks and leptons to Z^0 is given by Eq. (70), i.e., is proportional to the eigenvalue of $(T_3 - Q \sin^2\theta_W)$ for the fermion in question; see §VI.B.3, Vol. II.

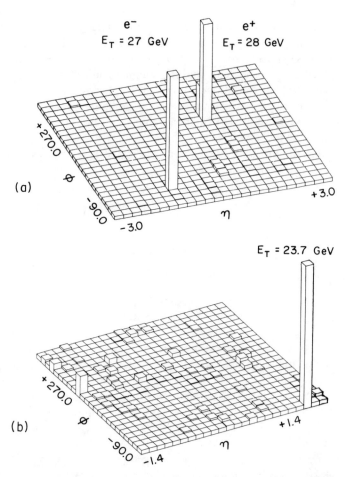

FIG. 37. Observation of events interpreted as due to the production and decay of vector bosons that mediate the weak interaction. These results are from the UA1 Collaboration's detector at the CERN SPS $p\bar{p}$ collider operating at a total c.o.m. energy of 540 GeV. Similar results have been reported by the UA2 Collaboration (1983), whose detector was already described in Fig. 28. The UA1 detector can discriminate between electrons, muons, and hadrons, and it is in a magnetic field, so it also determines the sign of the charge. Events due to the mechanisms of Eq. (77) are selected from a very large background of purely hadronic collisions by requiring one or more charged leptons with a large component of momentum transverse to the $p\bar{p}$ direction.

Events with two oppositely charged, high-momentum leptons are candidates for Z^0 production. Such events are ascribed to Z^0 if the leptons' transverse momenta balance to within an amount consistent with that carried off by the low energy hadrons that are also produced. An event of this type [from UA1 Collaboration (1983b)] is shown in (a). The coordinate mesh (φ, η) in this histogram represents the angular segmentation of the detector, which covers virtually the whole solid angle surrounding the interaction region. The $p\bar{p}$ direction is taken as the polar axis, while φ is the azimuth, and $\eta = \ln \tan \frac{1}{2}\theta$, where θ is the polar angle. Each detector cell measures the energy E deposited in it. The vertical columns show the quantity $E_T = E \sin \theta$, which equals the transverse momentum p_T for relativistic particles. The event shown in (a) is seen to have a highly energetic back-to-back e^+e^- pair, of which the positive member has traversed two cells,

146

The measured value of θ_W allows us to predict the masses of the intermediate bosons. From (71), (73), and (75) one finds

$$m_W = 83 \pm 3 \text{ GeV} \qquad m_Z = 94 \pm 2 \text{ GeV} \qquad (76)$$

These large masses indicate the magnitude of the symmetry breaking in the electroweak theory. The masses of W and Z^0 are almost as large as that of an atom of silver! Only at energy exchanges much larger than m_W and m_Z is the underlying symmetry apparent to the uninstructed observer. But we have exploited weak isospin invariance in the low-energy regime, where the symmetry is hidden, to predict the detailed structure of Z^0-mediated weak interactions. As we have seen, independent measurements give the same value of the basic parameter θ_W. In §VI.B.3 (Vol. II) we shall learn that more detailed properties of Z^0-mediated processes are also in accord with the basic Z^0-transmutation law, Eq. (70).

That several different experiments give the same value of θ_W, and are consistent with Eq. (72), constitutes an impressive success of the electroweak theory. But in all these "low-energy" experiments, the propagation of the field quanta W^+, W^-, and Z^0 is not observable, and a static approximation analogous to the Biot–Savart law can describe virtually all the data.

A set of very striking events observed at CERN in $p\bar{p}$ collisions has, however, given the first *direct* evidence for the existence of these field quanta (UA1 Collaboration, 1983a & b; UA2 Collaboration, 1983). These events are, presumably, due to the annihilation of a quark in p with an antiquark in $\bar{p}$, and the subsequent formation of $W^\pm$ or Z^0, which then decays leptonically:

$$u\bar{d} \to W^+ \to \bar{l}\nu_e, \ \bar{u}d \to W^- \to l\bar{\nu}_l, \text{ and } q\bar{q} \to Z^0 \to \bar{l}l. \qquad (77)$$

The charged leptons l and/or $\bar{l}$ are observed as highly energetic tracks, accompanied by low-energy hadronic debris produced by the four quarks and antiquarks that are spectators in the collision (see Fig. 37). The initial observations cited above, when combined with more recent data (see Appendix I, Table 1), are in excellent agreement with the theoretical predications of Eq. (76) and provide a striking confirmation of the electroweak theory. That the charged lepton in an event of the W-type [see Fig. 37(b)] is actually produced by the weak interaction is confirmed by the

---

accompanied by low energy particles. The Z^0 mass (and momentum) is determined directly from the e^+ and e^- momenta.

Events of the "W-type" require a more sophisticated analysis, because they supposedly have an energetic neutrino in the final state that is not detected. That is, events of this type have one lepton with large E_T, and a large (≥ 15 GeV) apparent violation of energy and momentum conservation. The histogram for such an event [from UA1 Collaboration (1983a)] is shown in (b). By definition, the W mass is found by ascribing the missing energy and momentum to a (massless) neutrino, and by fitting the charged leptons' observed momentum distribution with a formula derived from the theory sketched in the text.

observation (Spiro, 1983) of parity violation, in that such leptons display a fore-aft asymmetry with respect to the incident proton direction having a magnitude consistent with the theoretical expectation.*

A word of caution is in order, nevertheless. The electroweak theory supposedly incorporates quantum electrodynamics, and it must therefore pass very stringent tests if its validity is to be fully established. For example, the higher order corrections mentioned in connection with Eq. (73) will have to be confirmed in detail by high precision experiments. At this time the most promising technique for this purpose appears to be $e\bar{e} \to Z^0$, where Z^0 would then be observed to decay into $\bar{l}l$, or into hadrons. These matters are taken up in §VI.B.3, Vol. II.

11. Neutral kaons and CP violation

The neutral K-mesons give rise to a number of unique phenomena of great interest. On the one hand, some of these phenomena provide a particularly striking demonstration of the basic principle that a linear superposition of two state vectors is again a state of the system. On the other, a study of neutral K-decays shows that CP reflection, as defined in §9(f), is not an exact symmetry of nature.

(a) K^0–$\bar{K}^0$ mixing

As we learned in §3(b), there are two neutral kaons, K^0 and $\bar{K}^0$, with strangeness $S = 1$, and -1, respectively. They are the isospin partners of K^+ and K^-, respectively, and thereby form two distinct $I = \frac{1}{2}$ doublets (see Table 2), K and $\bar{K}$.

K^0 and $\bar{K}^0$ are stationary eigenstates only as long as one ignores the weak interaction H_{wk}, which does not conserve strangeness. Since H_{wk} is so feeble in comparison to the strong interaction, H_{st}, this is a highly accurate approximation that holds under almost all circumstances. But it does not hold for neutral kaons, and they owe their singular role in particle physics to this.

To understand why the neutral kaons are unusual, we first examine a general problem in quantum mechanics. Consider a Hamiltonian H that can be split into two parts, $H = H_0 + H_1$, with H_0 far larger than H_1. Let $(H_0 - \varepsilon_n)|\psi_n\rangle = 0$. As H_1 produces transitions among the $|\psi_n\rangle$, we can simplify the discussion by assuming that H_1 has only off-diagonal elements whose order of magnitude is M. If H_1 is to have an appreciable effect on any state $|\psi_n\rangle$, there must be at least one other state $|\psi_m\rangle$ that is nearly degenerate with $|\psi_n\rangle$. By "nearly degenerate" we mean that $\Delta = |\varepsilon_n - \varepsilon_m|$ must be of the order M, or smaller. If Δ is not zero, the perturbed nth eigenstate will have an admixture of $|\psi_m\rangle$ of the order M/Δ. When the

---

* The sign of this asymmetry can be deduced from Fig. 36, and its counterparts for $u\bar{d} \to W^+$, etc.

unperturbed states are degenerate ($\Delta = 0$), the perturbation H_1 will lift the degeneracy, and the new eigenvectors will be $(|\psi_n\rangle \pm |\psi_m\rangle)/\sqrt{2}$ if H_1 is Hermitian.*

In our problem $H_0 = H_{st}$, and $H_1 = H_{wk}$. The two states K^0 and $\bar{K}^0$ are degenerate eigenstates of H_{st} since they have the same mass if H_{wk} is neglected. But despite its weakness, H_{wk} cannot be neglected, because it does not conserve strangeness, and can therefore cause transitions between K^0 and $\bar{K}^0$. As a consequence, the true eigenstates of $H_{st} + H_{wk}$ are linear combinations of $|K^0\rangle$ and $|\bar{K}^0\rangle$. Virtually no other particle possesses these properties.** Consider, for example, the pairs Λ^0, $\bar{\Lambda}^0$, and K^+, K^-. In addition to strangeness, they also carry the exactly conserved quantum numbers $B = \pm 1$ and $Q = \pm 1$, respectively. Consequently, the strangeness-violating interaction H_{wk} cannot cause the transitions $\Lambda^0 \leftrightarrow \bar{\Lambda}^0$ or $K^+ \leftrightarrow K^-$ between these degenerate pairs of states.

Let us examine the K^0–$\bar{K}^0$ transition more closely. It corresponds to a change $\Delta S = 2$ in strangeness. The charge-changing weak interaction can only produce $\Delta S = \pm 1$, as is clear from Table 7, or Eq. (52). Nevertheless, the $\Delta S = 2$ transition K^0–$\bar{K}^0$ does occur as a two-step process. For example, if both K^0 and $\bar{K}^0$ can decay into some state χ, then $K^0 \leftrightarrow \chi \leftrightarrow \bar{K}^0$ will connect K^0 and $\bar{K}^0$. Examples of such states χ are $\pi^+\pi^-$ and $\pi^+\pi^-\pi^0$. In terms of quarks and W's, one can also depict the K^0–$\bar{K}^0$ mixing mechanism by Feynman diagrams, one example being

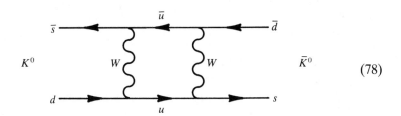

$$(78)$$

As we see, two W's are exchanged in a sequence of two $\Delta S = 1$ transitions. The neutral-current weak-interaction, as mediated by Z^0, cannot cause a $K^0 \leftrightarrow \bar{K}^0$ transition because this interaction conserves all flavors, as we shall demonstrate in §VI.B.3, Vol. II.

The eigenstates $|K_{1,2}\rangle$ of $H_{st} + H_{wk}$ are some linear combination of $|K^0\rangle$ and $|\bar{K}^0\rangle$. If CP is a symmetry of H_{wk}, as we shall assume for the time being, one can construct $|K_{1,2}\rangle$ without computation. Since $S \to -S$ under C, $C|K^0\rangle = |\bar{K}^0\rangle$. For free particles, the spatial wave functions of K^0 and $\bar{K}^0$ are

---

* In this last connection, see the footnote on p. 153.
** An exception are the charmed mesons D^0 and D^0. Since charm is not conserved by H_{wk}, D^0–$\bar{D}^0$ mixing can occur, but has not been observed thus far. The same remark applies to the $Q = 0$ mesons B^0 and $\bar{B}^0$ containing a b or $\bar{b}$ quark.

identical, so P has the same effect on both. We therefore have*

$$CP|K^0\rangle = |\bar{K}^0\rangle,$$
$$CP|\bar{K}^0\rangle = |K^0\rangle. \tag{79}$$

If CP is a symmetry of H_{wk}, $|K_{1,2}\rangle$ must be eigenstates of CP. In view of (79) these eigenstates are

$$|K_1\rangle = \frac{1}{\sqrt{2}}\,(|K^0\rangle + |\bar{K}^0\rangle),$$
$$|K_2\rangle = \frac{1}{\sqrt{2}}\,(|K^0\rangle - |\bar{K}^0\rangle). \tag{80}$$

The masses m_1 and m_2 of these states differ by a very small amount (see Appendix I.2).

As seen from (79), $|K_1\rangle$ is even, and $|K_2\rangle$ is odd under CP:

$$CP|K_1\rangle = |K_1\rangle,$$
$$CP|K_2\rangle = -|K_2\rangle. \tag{81}$$

Our provisional hypothesis that CP is a symmetry of H_{wk} then implies that the weak interaction will cause $|K_1\rangle$ and $|K_2\rangle$ to decay into final states that are, respectively, even and odd under CP.

Consider the final states $\pi\pi$ and $\pi\pi\pi$. Since kaons have $J = 0$, the $\pi\pi$ state has vanishing orbital angular momentum, and even parity, because its parity is the product of the spatial parity $+1$, and two factors of -1 for the intrinsic parity of each pion. In the case of $\pi^+\pi^-$, C interchanges the two pions, but this leaves an S-state unaltered. As a result, $\pi\pi$ is even under CP, and *only* K_1 can decay into it. As we shall show later, $\pi\pi\pi$ is odd under CP, but that is not important at the moment, because $3m_\pi \simeq 420$ MeV, which is rather close to $m_{K^0} \simeq 498$ MeV, while $2m_\pi \simeq 280$ MeV, which is not. The small energy release in 3π compared to 2π-decay means that a state that can undergo 2π-decay will have a far shorter lifetime than a state that is forbidden to decay in this way. From this we conclude that K_1, which can decay into 2π, will have a lifetime τ_1 that is much smaller than the lifetime τ_2 of K_2.

This difference of lifetimes has remarkable consequences. Consider a neutral kaon produced in some strong reaction, say in association with Λ^0 in a $\pi - p$ collision. Since H_{st} conserves S, any state produced in conjunction with Λ^0 must have the eigenvalue $S = 1$, and is therefore $|K^0\rangle$. But $|K^0\rangle$ is not an eigenstate of $H_{\text{st}} + H_{\text{wk}}$, so it has no well-defined mass or lifetime. To determine its time evolution, we must decompose $|K^0\rangle$ into the eigenstates

---

*Or, to be precise, we are free to adopt a phase convention such that (79) holds.

$|K_{1,2}\rangle$ of $H_{st} + H_{wk}$, which have the simple time dependence $e^{-im_k t} e^{-\frac{1}{2}t/\tau_k}$, where m_k and τ_k are the mass and lifetime of $|K_k\rangle$. The time dependence* of the K^0-state is then given by substitution into (73):

$$|K^0(t)\rangle = \frac{1}{\sqrt{2}} \{\exp(-im_1 t) \exp(-\tfrac{1}{2}t/\tau_1)|K_1\rangle$$
$$+ \exp(-im_2 t) \exp(-\tfrac{1}{2}t/\tau_2)|K_2\rangle\}, \qquad (82)$$

This equals $|K^0\rangle$ at the time of production, $t = 0$. At times t satisfying $t \lesssim \tau_1$, $|K_1\rangle$ and $|K_2\rangle$ are both present in $|K^0(t)\rangle$, and one will observe the $\pi\pi$ decay mode stemming from $|K_1\rangle$. But for $t \gg \tau_1$, the rapidly decaying component $|K_1\rangle$ will have disappeared, and only $|K_2\rangle$, which cannot go into 2π, will remain. To summarize, if one observes the 2π- and 3π-decay modes following K^0 production, one should see two quite different exponential decays: a rapidly decreasing number of 2π-decays, and a much more slowly decreasing number of 3π-decays. If the kaons move with high velocity, 3π-decays will be observed at much larger distances from the K^0-production region than 2π-decays.

This phenomenon has been studied in great detail. The short lifetime τ_1 is 0.89×10^{-10} sec, whereas the long lifetime τ_2 is 0.52×10^{-7} sec. The 2π-decay mode accounts for $\sim 100\%$ of the K_1-decays, whereas 3π-decay provides $\sim 34\%$ of the K_2-decays. In the latter case some 65% of the decays go into the semileptonic modes $\pi^{\pm} l^{\mp} \nu$, where l is e or μ, and ν the neutrino dictated by conservation of N_l. That $\tau_2 \sim 10^3 \tau_1$ is consistent with what one expects from 3π- vs. 2π-decays because of the large difference in available phase space.**

Another interesting effect comes to light if one asks how a neutral kaon beam interacts with hadrons, say a hydrogen target. The resulting reactions are very sensitive to whether the collision involved an $S = +1$ or $S = -1$ projectile, because there are many $S = -1$ baryons that can be formed in a Kp collision, whereas there are no baryons with $S = 1$ (see Table 1). But the admixture of $S = 1$ and $S = -1$ in (82) changes with time! It is straight-forward exercise to verify that Eqs. (82) and (80) lead to the conclusion that at time t the probability of finding K^0 in $|K^0(t)\rangle$ is

$$P(K^0,t) = \frac{1}{4}\left[\exp(-t/\tau_1) + \exp(-t/\tau_2)\right.$$
$$\left. + 2 \exp\left[-\frac{1}{2}\left(\frac{1}{\tau_1} + \frac{1}{\tau_2}\right)t\right]\cos(t\Delta m)\right], \qquad (83)$$

---

* This is the time dependence in the rest frame, where the energy is equal to the mass. Further discussion of the exponential decay factor will be found in §III.A.3, Vol. II. If the particles move with velocity v, one must make the substitutions $m \to \gamma m$, $\tau \to \gamma\tau$, where $\gamma = (1 - v^2)^{-}$.

** Semileptonic decay of K_1 is also possible, but is very rare in comparison to the dominant 2π mode.

where $\Delta m = |m_1 - m_2|$ is the $K_1 - K_2$ mass difference. If the kaon beam is fairly monoenergetic, one should therefore see a variation in $S = 1$ vs. $S = -1$ final states as one changes the distance between the secondary target, and the K^0 production point. Because of the oscillating factor $\cos t\Delta m$ in (83), this permits a measurement of Δm, with the result that $\Delta m = 3.52 \times 10^{-6}$ eV. This remarkably small mass difference is of the magnitude that one expects from the mechanism depicted in Eq. (78).

The considerable difference between KN and $\bar{K}N$ interactions to which we have just referred leads to *the regeneration phenomenon*. Consider, for the sake of concreteness, a pure K_2-beam, which is what one has if one goes far from the production region. We then examine this beam after it has passed through a slab of matter. Because K^0 and $\bar{K}^0$ have quite different interactions with nucleons, and are not mixed by the strong interaction, it is convenient to decompose $|K_2\rangle$ into $|K^0\rangle$ and $|\bar{K}^0\rangle$ before it enters the slab. After passing through the slab, these become $T|K^0\rangle$ and $\bar{T}|\bar{K}^0\rangle$, where T and $\bar{T}$ are quite different transmission amplitudes. As a consequence, the transmitted beam is*

$$|K_2\rangle_{\text{trans}} = \frac{1}{\sqrt{2}} [T|K^0\rangle - \bar{T}|\bar{K}^0\rangle]$$

$$= \frac{1}{2} [(T - \bar{T})|K_1\rangle + (T + \bar{T})|K_2\rangle], \qquad (84)$$

where (80) and its converse were used. As we see, the short-lived K_1 component has been regenerated, and is readily observed by detecting copious 2π-decays on the downstream side of the slab far from the production points, whereas there are no** 2π-decays on the upstream side of the slab.†

(b) CP violation

Our first task is to examine the *CP* transformation of the three-pion decay mode of the neutral kaons. We begin with $\pi^0\pi^0\pi^0$; the analysis of $\pi^+\pi^0\pi^-$ is similar, though a bit more complicated, and leads to the same result. Let **L** be the orbital angular momentum of two π^0's in their c.o.m. system, and **l** the

---

* We assume the slab is thin enough to allow us to ignore the time dependence due to decay, and the $K_1 - K_2$ mass difference.

**Here we ignore a small ($\sim 10^{-3}$) contamination of 2π-decays due to *CP*-violation.

† All the effects we have just discussed have analogies in optics. Consider a slab of material in a magnetic field, and photons travelling along the field direction. The index of refraction and absorption coefficient for left and right circularly polarized photons, $|L\rangle$ and $|R\rangle$, will differ, and there will be no mixing between $|L\rangle$ and $|R\rangle$. Assume that the absorption coefficient for $|R\rangle$ is much larger than that for $|L\rangle$, and allow one of two linearly polarized waves, $|x\rangle$ or $|y\rangle$, to fall on the slab. What then occurs is precisely analogous to the neutral kaon phenomena if one makes the correspondence $|x\rangle \to |K^0\rangle$, $|y\rangle \to |\bar{K}^0\rangle$, $|L\rangle \to |K_2\rangle$, $|R\rangle \to |K_1\rangle$. If only $|L\rangle$ survives after transmission through the slab, another thin slab in which the magnetic field is oriented along a different direction will "regenerate" $|R\rangle$.

orbital angular momentum of the third π^0 with respect to the c.o.m. of the other two. Then $\mathbf{J} = \mathbf{L} + \mathbf{l}$; but kaons have no spin, so $J = 0$, and therefore $L = l$. Any two π^0's must be in a symmetric state because they are identical bosons, whence $L = 0, 2, \ldots$, and therefore the final state is $L = l = 0$, $2, \ldots$. All such spatial wave functions are even under space reflection; furthermore, C leaves all π^0's unaltered. As a result, the CP eigenvalue of a $J = 0, 3\pi^0$ state is $(-1)^3 = -1$, because each π^0 has odd intrinsic parity. This establishes our earlier claim that the 3π-decay mode is odd under CP.

If CP is conserved, neither of the eigenstates $|K_1\rangle$ and $|K_2\rangle$ can decay into *both 2π and 3π*. Refined observations show that the long-lived component of a neutral kaon beam decays into *both 2π and 3π*. Consequently this long-lived state $|K_L\rangle$ cannot be $|K_2\rangle$, but must be a mixture of the CP-even and CP-odd states.* The observed branching fractions of $|K_L\rangle$ into the CP-even 2π-states are

$$
\begin{aligned}
K_L \to \pi^+\pi^- &: (2.03 \pm 0.05) \times 10^{-3} \\
\to \pi^0\pi^0 &: (0.94 \pm 0.18) \times 10^{-3}.
\end{aligned} \tag{85}
$$

Hence CP violation is a very small effect which, at this time, has only revealed itself in the remarkably sensitive neutral kaon system.

An even more graphic demonstration of CP violation is provided by the semileptonic decays of K_L. Consider the final states $X = \pi^+ e^- \bar{\nu}_e$ and $\bar{X} = \pi^- e^+ \nu_e$. Under CP, $X \leftrightarrow \bar{X}$, so if CP were a symmetry, and the initial state an eigenstate of CP, the rates for decay into X and $\bar{X}$ would have to be equal. The experimental facts (see Fig. 38) are

$$
\frac{\text{Rate}(K_L \to \pi^- l^+ \nu) - \text{Rate}(K_L \to \pi^+ l^- \bar{\nu})}{\text{Rate}(K_L \to \pi^- l^+ \nu) + \text{Rate}(K_L \to \pi^+ l^- \bar{\nu})} = (3.3 \pm 0.1) \times 10^{-3}, \tag{86}
$$

where l is e and μ. In other words, the object K_L prefers to decay into positrons rather than electrons, and thereby provides a clear-cut distinction between matter and antimatter. This allows us to give a convention-independent definition of positive charge: it is the charge carried by the preferentially emitted spin $\frac{1}{2}$ particles stemming from the long-lived component of a neutral kaon beam.

---

* The long-lived and short-lived eigenstates are

$$
|K_L\rangle = (1 + |\varepsilon|^2)^{-\frac{1}{2}}[|K_2\rangle + \varepsilon|K_1\rangle],
$$
$$
|K_S\rangle = (1 + |\varepsilon|^2)^{-\frac{1}{2}}[|K_1\rangle + \varepsilon|K_2\rangle],
$$

which reduce to $|K_2\rangle$ and $|K_1\rangle$ if $\varepsilon = 0$. The CP violation parameter ε is defined as $\varepsilon = (2\eta_{+-} + \eta_{00})/3$, where $\eta_{+-} = A(K_L \to \pi^+\pi^-)/A(K_S \to \pi^+\pi^-)$, and η_{00} is the same ratio of decay amplitudes for the $\pi^0\pi^0$ mode. The measured value is $\varepsilon \approx 1.6(1 + i) \times 10^{-3}$. Incidentally, there is no typographical error: $|K_L\rangle$ and $|K_S\rangle$ are not orthogonal if $\varepsilon \neq 0$. The effective Hamiltonian that describes the neutral kaons is not Hermitian because it describes their decay into other states, and its eigenvectors are only orthogonal under certain conditions which are not met if CP is violated. For a detailed treatment of neutral kaons, see Lee and Wu (1966); Gottfried (1970), §3.6; and Lee (1981), §15.3.

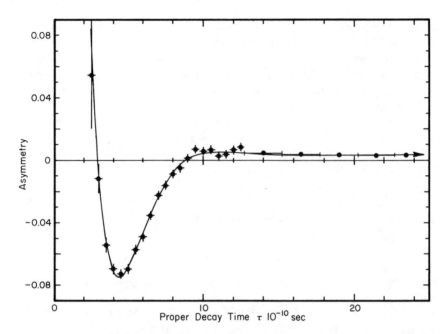

FIG. 38. *CP* violation in $K^0 \to \pi e \nu_e$ decay, and oscillations due to the $K_L - K_S$ mass difference, Δm. This experiment (Gjesdal, 1974) measures the asymmetry, defined as in Eq. (86), as a function of the elapsed proper time τ between K^0 production and decay.

Energy and momentum conservation, together with the known direction of the kaon beam, suffice to determine the K^0 momentum, and therefore the relationship between proper time and the distance between the production point and the decay vertex. The oscillatory behavior of the asymmetry arises because $|K^0(t)\rangle$ is a superposition of $|K_L\rangle$ and $|K_S\rangle$, having a form similar to Eq. (82), though somewhat more complicated because of *CP* violation. To a good approximation, the asymmetry as a function of τ is given by

$$2(1 - |x|^2)|1 - x|^{-2}[\text{Re } \varepsilon + \exp(-\tau/\tau_L) \cos(\tau \Delta m)],$$

where ε is defined in the footnote on p. 153, and x is the ratio of amplitudes for K^0-versus $\bar{K}^0$-decay into the final state $\pi^- e^+ \nu_e$. Among other things, these data determine the mass difference Δm quoted in Appendix I.2.

At this time it is not known whether *CP* violation is due to the conventional weak interaction, or to a distinct interaction.

12. Synopsis: the Standard Model

This volume has presented an overview of the concepts of particle physics. It began with concepts that prevailed when modern physics was born some three centuries ago, and ended with concepts that recent research has firmly established. Our purpose in this section is to summarize this newly found orthodoxy, which is frequently called the Standard Model. The last section of this volume will venture into territory that is still largely uncharted.

(a) The Quantum Ladder

The internal structure of matter reveals a most striking feature which we shall call the Quantum Ladder. Each rung of the Ladder pertains to phenomena that have quite unique properties, and therefore appears as a distinct branch of physics. A rung is distinguished from its neighbors by a dramatic difference in the order of magnitude of the energy transfers, and therefore in the dimensions of the motions involved.

Three rungs on the Quantum Ladder are now firmly established:*
1. The atomic rung, characterized by energy exchanges ΔE_c of order eV, and characteristic dimensions R_c of order 10^{-8} cm;
2. The nuclear rung, where $\Delta E_c \sim 10^6$ eV, and $R_c \sim 10^{-12}$ cm; and
3. The subnuclear rung, with $\Delta E_c \sim 10^9$ eV, and $R_c \sim 10^{-13} - 10^{-14}$ cm.

Within each rung there is a rich spectroscopy whose level spacings are of order ΔE_c.

The existence of the Quantum Ladder has a profound influence: degrees of freedom that are active in one rung are frozen in the rungs below. This is so because systems whose internal motions provide the excitation spectrum within a given rung are in their ground states when the energy available confines observation to a lower rung. Objects that obviously have internal structure when viewed from the vantage point of the higher rung therefore appear as structureless in all phenomena on any lower rung. Hence the atom is an elementary particle in gas kinetics, the nucleus in atomic physics, and so on.

In view of this, it is natural to ask whether there is such a thing as absolute elementarity—whether the Ladder has infinitely many rungs. No known principle requires that. It is quite possible that the Ladder is finite, and that some of the objects that we now call elementary particles have no substructure whatsoever. Obviously such a contention could never be established purely by experiment. But one can imagine the existence of a theory that involves structureless objects, and whose predictions are so logically compelling and empirically successful, as to engender confidence that those objects are, indeed, elementary. The present state of theoretical physics is so far from such an ideal and complete theory that there is no justification for dwelling on the implications of such a possibility.

While bearing in mind that further rungs on the Ladder may exist, we must take stock of those objects that are today's candidates for ultimate elementarity. They fall into two categories, spin $\frac{1}{2}$ fermions and spin 1 bosons.

(b) Fermions

We have learned that there are two quite distinct types of "fundamental" fermions: leptons and quarks. The quarks carry the tri-valued color quantum number, whereas the leptons do not. At first sight it is natural to separate

---

* One could also define a zeroth, or classical, rung, wherein the energies are too low to produce any excitations in atoms or molecules.

these fermions into three families as follows:

$$\begin{bmatrix} u & \nu_e \\ d & e \end{bmatrix} \quad \begin{bmatrix} c & \nu_\mu \\ s & \mu \end{bmatrix} \quad \begin{bmatrix} (t) & \nu_\tau \\ b & \tau \end{bmatrix}. \tag{87}$$

Each family contains two quarks and two leptons. The quarks have charges $Q = \frac{2}{3}$ and $Q = -\frac{1}{3}$, the leptons $Q = 0$ and $Q = -1$. Since the quarks are threefold degenerate, the total charge of each family vanishes. As one moves from left to right in (87) the masses increase by sizeable steps for all particles, with the possible exception of the neutrinos, which could all have zero mass. At this time there is no direct experimental evidence for the t-quark, and for that reason it is shown in parentheses in (87). But there is good reason to anticipate its existence. From a naive point of view, the pattern shown by Eq. (87) cries out for a $Q = \frac{2}{3}$ partner to b, but there is also a more compelling argument in its favor.* From the data on e^+e^- annihilation now available, the limit on the mass of t is $m_t \gtrsim 20$ GeV.

The first family in (87) plays a special role, as virtually everything we see about us in nature only involves fermions in this family. With the possible exception of the neutrinos, the members of the second and third families are unstable. Hence they only appear in accelerator laboratories and in cosmic ray phenomena; as we shall see in §13, they also are presumed to have played an important part in the very early history of the universe, immediately following the Big Bang.

Both quarks and leptons also carry the quantum number called weak isospin T, and all belong to $T = \frac{1}{2}$ doublets. Weak isospin has deep significance because it reflects a hidden symmetry of the electroweak interaction that is exact insofar as we now know.

The leptons within *each* family belong to *one* such T-doublet, but the situation is somewhat more involved for the quarks. This is most readily seen by recalling [§9(b)] that three lepton numbers N_l are conserved by all the known interactions, so that there are no $\mu \leftrightarrow e$ transitions, etc. On the other hand, there are no corresponding conservation laws for quarks of different flavor; for example, it is not true that

$$N(u) - N(\bar{u}) + N(d) - N(\bar{d})$$

is conserved, because there are $s \leftrightarrow u$ transitions, etc. But one can form linear combinations of the $Q = -\frac{1}{3}$ quarks (or, equivalently, of the $Q = \frac{2}{3}$ quarks) so that the charge-changing weak interaction only produces transitions between *one* $Q = \frac{2}{3}$ quark state and *one* $Q = -\frac{1}{3}$ state. If there were but two families, the appropriate linear combinations of $Q = -\frac{1}{3}$ quarks, d' and s',

---

* The weak decays of the B-mesons (with composition $b\bar{u}$ and $b\bar{d}$) reveal that the b-quark decays predominantly into c, and not u. It can be shown that this is to be expected if b is a member of a weak isodoublet with a sixth quark (t), but is inconsistent with models having only five quarks.

were already given explicitly in Eq. (48). When there are three families, there are three such "eigenmodes" of the charge-changing weak interaction, d', s', and b', and the coefficients that appear in the analogue of Eq. (48) depend not only on the Cabibbo angle, but also on several other parameters. The quark T-doublets are then (u,d'), (c,s'), and (t,b'). When we wish to emphasize the role of weak isospin it is therefore not appropriate to group the fundamental fermions as in Eq. (87), but rather as follows:

$$
\begin{bmatrix} u & \nu_e \\ d' & e \end{bmatrix}
\begin{bmatrix} c & \nu_\mu \\ s' & \mu \end{bmatrix}
\begin{bmatrix} (t) & \nu_\tau \\ b' & \tau \end{bmatrix}.
\tag{88}
$$

Henceforth we shall refer to these groupings as *generations*.

It is illuminating to represent the fermions within each generation in a three-dimensional Cartesian coordinate system $\mathfrak{C}$ that displays their quantum numbers (see Fig. 39). One axis, say the z-axis, specifies the component T_3 of the weak isospin. As all fermions have $T_3 = \frac{1}{2}$ or $-\frac{1}{2}$, they must all appear on one or the other of two planes perpendicular to the z-axis with intercepts at $z = \pm\frac{1}{2}$. The two other axes, called x and y, are used to specify color, and any object without color lies at $x = y = 0$. Hence the leptons are placed on the z-axis, as shown. The three color states of each quark are placed on the vertices of an equilateral triangle in the plane having the T_3-intercept

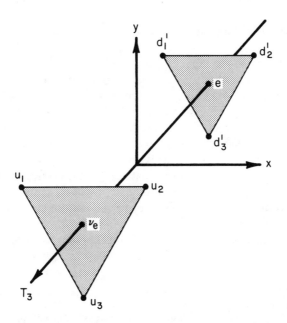

Fig. 39. The leptons and quarks belonging to the first generation, shown in the coordinate system $\mathfrak{C}$, whose z-axis specifies the component T_3 of weak isospin, and whose x and y- axes specify the color quantum numbers; for a detailed discussion of the latter, see §IV.A (Vol. II).

appropriate to that quark. The complete symmetry between the three color states is captured by the equilateral nature of the triangles; the mathematical reasoning behind this assertion will be explained in §IV.A.3 (Vol. II).

We emphasize that each fermion generation has a diagram identical to Fig. 39.

(c) Vector bosons and gauge fields

The strong and electroweak interactions are mediated by fields whose quanta have spin 1, and are therefore vector bosons. These are the basic elementary bosons of the Standard Model.*

The most familiar field quantum is the photon, γ. It can be emitted or absorbed by any fermion with nonzero charge, and in so doing the fermion does not alter its color or weak isospin eigenvalue T_3. Hence a fermion stays put in Fig. 39 when it participates in the electromagnetic interaction.

The emission or absorption of other field quanta can, however, cause a change of color or T_3, and it is instructive to follow these changes in the space $\mathfrak{C}$ of Fig. 39. These emission and absorption processes may be separated into three categories:

1. Emission or absorption of a W^+ or W^-, which causes a quark or lepton to change its T_3-eigenvalue, and therefore produces a motion parallel to the z-axis;

2. Emission or absorption of gluons that cause a quark to change its color, and therefore to move along an edge of the triangle to which it belongs; and

3. Emission or absorption of field quanta that cause no change of color or T_3.

Let us count the number of independent emission and absorption processes, and thereby, the total number of field quanta. In §7(b) we learned that there are eight distinct gluons; six correspond to the possible motions to and fro along the edges of the color triangle, and two to field emission or absorption processes wherein the quark color does not change. To this we must add the four electroweak quanta: W^+, W^-, and the two quanta that cause no change of T_3: Z^0 and γ. This gives a total of twelve field quanta. We can also depict these quanta in the space $\mathfrak{C}$, as shown in Fig. 40, where the color and T_3-transitions just described are shown as arrows, while the four quanta that produce no change in the fermion quantum numbers are shown as dots at the origin of $\mathfrak{C}$.

Each of these twelve vector bosons is the quantum of a vector field. The equations of motion that determine the space-time behavior of these fields

---

* As we shall see, the gravitational interaction can be ignored in elementary particle physics as long as the total energy is below 10^{19} GeV. For that reason we shall not discuss the graviton, the spin 2 particle that emerges from a quantization of Einstein's field equations. The simplest models of the electroweak symmetry breaking also require a spin 0 Higgs particle [recall §10(c)], but as there are theoretical reasons to doubt that this can be an elementary particle [see §13(d)], we shall not include it in our fundamental boson zoo.

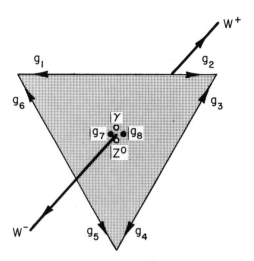

FIG. 40. The vector bosons of the Standard Model, represented in the coordinate system $\mathfrak{C}$ by the transitions that they mediate. The arrows in the x–y plane correspond to the six color-changing gluons, while the solid dots represent the two gluons that do not induce color changes. The charged W's produce changes of T_3, as indicated, while the T_3-conserving Z^0 and photon are shown as open circles.

will be discussed in Vol. II, Chaps. IV and VI. Here we shall confine ourselves to some descriptive remarks.

According to current theory, the strong and electroweak fields have equations of motion, and couplings to their sources, that are wholly specified by an elegant generalization of the principle of electromagnetic gauge invariance.* For a particle of charge Q, electromagnetic gauge invariance stipulates that the Schrödinger equation is to be invariant when the particle's wave function undergoes an arbitrary, space-time dependent, change of phase:

$$\psi(\mathbf{r},t) \rightarrow e^{-i\chi(\mathbf{r},t)}\,\psi(\mathbf{r},t). \tag{89}$$

The invariance of the kinetic energy term requires the introduction of a field, the vector potential $\mathbf{A}(\mathbf{r},t)$. It appears in the Schrödinger equation as an addition to the gradient, in the combination

$$\boldsymbol{\nabla} - iQ\mathbf{A}, \tag{90}$$

and undergoes the transformation

$$\mathbf{A} \rightarrow \mathbf{A} - \frac{1}{Q}\,\boldsymbol{\nabla}\chi \tag{91}$$

* Gauge theories are discussed in Vol. II, Chaps. IV and VI; for an introductory survey, see Weisskopf (1981).

in concert with (89). Gauge invariance requires that observable quantities be independent of χ when the transformations (89) and (91) are carried out. For that reason the magnetic field is the curl of $\mathbf{A}$, and the electric field follows therefrom by a Lorentz-covariant generalization of the curl. The replacement of the gradient in the kinetic energy by (90) gives rise, in the energy density, to a term $\mathbf{j} \cdot \mathbf{A}$ that couples the field to the electromagnetic current density $\mathbf{j}$. This is the origin of the electromagnetic interaction, as given by Eq. C(9).

The electroweak and color fields are introduced in a similar fashion. In the former case, the fermion sources form a $T = \frac{1}{2}$ doublet, and instead of (89) their wave function is therefore transformed by a 2×2 matrix U_2. Aside from an overall phase, such a matrix depends on three real parameters, as we learned in §B.3(d). This unitary transformation is elevated to the status of a (non-Abelian) gauge transformation by letting these three parameters be arbitrary space-time dependent functions. One of these functions specifies rotations generated by the diagonal Pauli matrix τ_3, and is therefore not involved in any process that changes T_3. The other two functions parametrize rotations about the 1 and 2 axes in the weak isospin space $\mathscr{E}_3^T$, and are associated with changes of T_3. The generalization of gauge invariance to the 2×2 case therefore requires the introduction of *three* vector potentials, one per parameter. One is connected to transformations generated by τ_3, its quantum is called W^0, and its emission or absorption causes no change of T_3. The other two fields, connected with $\tau_1 \pm i\tau_2$, cause transitions between the $T_3 = \frac{1}{2}$ and $-\frac{1}{2}$ states of the fermions; their quanta are W^+ and W^-. There is a fourth field, associated with the overall phase mentioned at the beginning of this paragraph. Its quanta are the objects called B^0 in §10(b). The electrically neutral quanta B^0 and W^0 combine to form the photon and Z^0, as explained in the discussion leading to Eqs. (64) and (66).

The color fields arise when one demands that a transformation on a quark triplet induced by a 3×3 unitary matrix U_3, whose elements are arbitrary functions of space and time, leave the quarks' Schrödinger equation invariant. Aside from an overall phase, which was already used for the B^0 field, U_3 has eight other free parameters. To each of these there corresponds a color field, and a gluon field quantum, as we already saw in connection with the discussion related to Fig. 40.

Because of the close parallel between the electroweak and color gauge invariance principles on the one hand, and electromagnetic gauge invariance on the other, all couplings have the electromagnetic format of Eq. C(9): the scalar product of a current and a vector potential. There are 12 distinct currents, one per gauge field. These currents describe the transport of color, weak isospin, and naturally also of electrical charge. The field equations also bear a striking resemblance to Maxwell's equations.*

---

* Indeed, by introducing a suitable notation, one can write the electroweak and strong field equations (the so-called Yang–Mills equations) in a format identical to Maxwell's; see §IV.C.3 (Vol. II).

Despite these similarities between pure electrodynamics and the strong and electroweak interactions, there are very significant differences. As we learned in §7(b), QCD is an intrinsically nonlinear field theory because the gluons transport color. The same is true of the electroweak theory. These nonlinearities manifest themselves in the form of transitions such as $W^+W^- \leftrightarrow Z^0$ and $W^+W^- \leftrightarrow Z^0Z^0$ among the field quanta, and to corresponding vertices in the Feynman diagrams. In the case of the color field, these vertices were already shown in Fig. 20(b) and (c).

This completes our summary of the Standard Model—that is, of what we believe to be either established by direct observation, or by compelling theoretical inference. This orthodoxy could still be overthrown, because many of its essential ingredients have not yet been seen in the laboratory: The t-quark, the mutual couplings of the electroweak quanta, and the couplings between gluons; furthermore, the evidence for gluons from jet phenomena is still fragmentary. But let us be presumptuous enough to assume that this imposing list of gaps will be filled in the anticipated manner. Then we can say that all the phenomena of atomic, nuclear, and subnuclear physics observed thus far are ultimately due to the fermions listed in Eq. (88), and the bosons shown in Fig. 40. By "thus far" we allude to existing data which show that the charged leptons, and the u and d quarks, are pointlike to within the currently available resolution of order 10^{-16} cm (see Chaps. II and V, Vol. II). If there is any lepton or quark substructure, it must have dimensions smaller than that. The discovery of such a substructure would herald the existence of at least one new rung on the Quantum Ladder, with characteristic energies of several hundred GeV or more.

13. Outlook

We wish to leave the reader with a glimpse of those conjectures that flow most plausibly from what we now know, but which have little, if any, experimental confirmation.* With this end in mind, we pose a number of questions that immediately arise from the foregoing synopsis:

1. Quarks and leptons are all structureless spin $\frac{1}{2}$ particles, having identical electroweak interactions, and appearing in an identical manner in each generation. Does this imply that quarks and leptons are related to each other?

2. The electroweak connection shows that electromagnetic and weak phenomena are actually manifestations of one interrelated set of fields. Is the strong interaction field also related to the electroweak field?

---

* There are many introductory articles on the topics to be touched on here. Among these we may mention the following: on grand unification, Georgi (1980, 1981) and Salam (1980); on proton decay, cosmology, and related matters, Goldhaber (1980), Schramm (1983), Turner (1979), Weinberg (1977, 1980), Weisskopf (1983), and Wilczek (1980); on monopoles, Carrigan (1982); on quark and lepton substructure, Harari (1982).

3. Why do the fermions occur in distinct generations that appear to be identical replicas of each other?

4. Why have we ignored *CP* violation in our synopsis?

The first question can be rephrased in the following manner:

1'. Are the quarks and leptons within one generation merely different states of a fermion multiplet, just as the charged lepton and the corresponding neutrino are members of a weak isospin doublet?

When put this way, the question becomes more incisive, because the notion "multiplet" only has content if transitions between states belonging to the multiplet are possible. In the present context these would be processes where a quark turns into a lepton, or vice versa. The existence of such transmutations would have dramatic implications, for they violate both baryon and lepton number conservation. Consequently, proton decay would no longer be forbidden: processes such as $p \to \pi^0 e^+$ would be allowed. Since the proton is known to have a lifetime that is vastly longer than the age of the universe, the proposal to put quarks and leptons into one multiplet is, at least at first sight, a rather far-fetched idea.

(a) Grand Unification

Let us now look at the second question. It asks whether all the fields of "elementary" particle physics can be combined into one "Grand Unified Field." If we assume that the field concepts that we have relied on thus far are generally valid, the conjecture that quarks and leptons are to be combined into multiplets has very direct implications for the nature of this Grand Unified Field. This is so because a transition between two members of a source multiplet must then be accompanied by the emission or absorption of a spin 1 field quantum. As shown in Fig. 40, in the Standard Model all transitions are either along the weak isospin axis T_3 of the coordinate frame $\mathfrak{C}$, or in one of the color planes orthogonal thereto, a circumstance which reflects the absence of any relationship between the electroweak and strong fields in that model. But a quark–lepton transition is along a direction that has both T_3- and color projections. For that reason the fields associated with the quark–lepton transitions would relate the strong to the electroweak fields.

Admittedly, the preceding geometric argument is no substitute for a mathematical theory. Nevertheless, it does reveal some of the essential features that a field theory based on lepton–quark unification must have: a set of fields that is larger than those of the Standard Model, with the new fields providing relationships between the electroweak and strong interactions. In particular, quantities that are free parameters in the Standard Model may have a definite value fixed by symmetry in the unified theory.

But how can a theory of this type be compatible with the lifetime τ_p of the proton, which is longer than 10^{31} years? This problem can, at least in principle, be overcome by simply copying the mechanism of the electroweak theory, which allows that theory to account for the great disparity between the rates of weak and electromagnetic processes at low energies. As we know, this is presumed to be due to the large masses of the W and Z^0 bosons, which inhibit weak processes at energies well below those masses. By the same token, one can make the proton lifetime arbitrarily long by simply assigning sufficiently large masses to the new bosons associated with the lepton–quark transitions. The order of magnitude of these masses can be estimated by the argument associated with Eqs. (41) and (42): Since, by hypothesis, the theory is to unify the interactions, we can take the electroweak coupling constant g to be the characteristic coupling constant for *all* interactions, including the one that is responsible for proton decay. The amplitude for proton decay must then be proportional to $(g/m_X)^2$, where m_X is the mass of a new, super-heavy field quantum. If we assume that τ_p is of order 10^{31} years, this leads to the estimate

$$m_X \sim 10^{15} \text{ GeV}. \tag{92}$$

This is a prodigious mass, which is just another way of saying that τ_p is enormous on the time scale set by the familiar processes of particle physics, and indeed of astrophysics.

The general features we have just sketched have been fleshed out in a number of concrete and detailed models. All interactions in these theories are mediated by gauge fields. Their predictions depend on the fact (see §IV.C.4, Vol. II) that in quantum field theories the amplitudes that describe the conversion of a field quantum into a fermion–antifermion pair, or vice versa, vary logarithmically with the total energy of the pair. This variation can be accounted for by replacing the coupling constant (e.g., e in QED) by a so-called *running coupling constant*, whose variation with energy is chosen to reproduce the aforementioned variation of the annihilation and creation amplitudes. This is more than a sleight-of-hand, because the scattering of two fermions has a variation with momentum transfer as if the basic interaction in the field theory were energy dependent, and this dependence is fully accounted for by replacing the elementary coupling constant by the running coupling constant. The rate at which the coupling constant "runs" is firmly predicted by the theory, and there is a characteristic difference between linear (or Abelian) field theories (such as QED) on the one hand, and intrinsically nonlinear (or non-Abelian*) field theories (such as QCD) on the other: In the linear case, the interaction becomes stronger with increasing energy, whereas in the nonlinear cases the interaction becomes weaker, and tends asymptotically to zero. The latter property of non-Abelian field theories is called *asymptotic freedom*.

The most constrained Grand Unified Models assume that all leptons and quarks within one generation form a multiplet, and that there are no missing

---

* This terminology has the following origin. A group is said to be Abelian if all its elements commute; otherwise it is non-Abelian. The elements of the "electromagnetic gauge group" are the phase factors in Eq. (89), which commute for differing values of χ, whereas the non-Abelian gauge groups involve noncommuting matrices, as discussed on p. 160.

fermions (except, possibly, for further complete generations). These models have only one fundamental coupling constant, to which the familiar coupling constants of the strong and electroweak interactions are related by the underlying symmetry. At energies currently available to us, this symmetry is not obvious—it is hidden, in the same sense that the electroweak symmetry is hidden at energies below $\sim$ 100 GeV [recall §10(c)]. But in view of the variation of the coupling constants discussed in the preceding paragraph, the strong interaction strength decreases as the energy of phenomena increases, whereas the electromagnetic coupling increases. Hence there is an energy M where these couplings become equal.* The quantity M can be computed from the gauge theories, and the values of the coupling constants at currently accessible energies, where they are very different indeed. Remarkably enough, the value of M calculated in this way from the Grand Unified Theories is of the same order of magnitude as our rough guess for the mass m_X. Furthermore, the calculation of the unification mass M automatically determines a well-defined value for the electroweak mixing angle θ_W.

The value of θ_W computed from the theory is in excellent numerical agreement with the latest measurements, as given in §10(d). The theoretically expected value for τ_p is somewhat *smaller* than the present experimental *lower* limit on the lifetime (Bionta, 1983). This disagreement between the expected and measured values of the proton lifetime, if confirmed, would, at the least, imply that the simplest Grand Unified Models are not valid. It may also be that the whole framework of grand unification is a misconception. Nevertheless, it is noteworthy that Grand Unified Theory predicts a proton lifetime that is enormously long in comparison to the characteristic time scales of particle physics, and a value of θ_W in astonishing agreement with the data. These successes indicate that the Grand Unified Theory should be treated with due respect.**

On the other hand, the Grand Unified Theories also have a number of very unsightly blemishes. They offer no insight whatsoever into the mass spectrum of the quarks and leptons. Indeed, these masses must be inserted into the theory "by hand." Nevertheless, the actual values of these masses play a crucial role in all the phenomena that surround us. The small electron-to-proton mass ratio causes the nuclei to have well-defined locations in the

---

* The situation is actually more complex (see Vol. II, §VI.D). As we learned in §10, the electroweak interaction has two independent coupling constants, g and g'. The former characterizes the coupling of the non-Abelian W field, and therefore decreases with energy, whereas the latter is associated with the Abelian B field, and increases with energy. Hence there are two falling coupling strengths, governing the W and gluon fields, and one growing strength, g'. Unification becomes manifest, as compared to hidden, at energies above the point where these three interactions merge, and the requirement that there be such a point determines not just M, but also θ_W.

** As shown in Vol. II, §VI.D, these theories actually predict θ_W and $\ln(m_X/m_W)$. Hence θ_W is insensitive to the details of the calculation, whereas τ_p, which is proportional to m_X^4, is very sensitive to the details of the theory.

surrounding electron cloud. This is the prerequisite for molecular architecture; without it, the stable configurations that characterize life and much of our familiar environment could not exist. The near-equality of the neutron and proton masses (i.e., of the u and d masses) is of fundamental importance in nuclear physics, and therefore in the sequence of processes that led from the early universe to the abundance of elements that we have today. Furthermore, Grand Unified Theories provide no hint as to why there are three generations of fermions. Insofar as the mathematical theory is concerned, one generation would do equally well. This leads one to ask whether a theory that has such a profound objective—the unification of quarks and leptons, and of all the known interactions of particle physics—can have so many adjustable parameters.

Leaving these doubts aside for the moment, we discuss some other consequences of grand unification. First, the theories predict the existence of *magnetic monopoles* having a mass comparable to that of m_X. Such objects were first proposed by Dirac a half-century ago, because the existence of a magnetic monopole guarantees that all electric charges are integer multiples of a fundamental unit of charge, and therefore in accord with what we observe. The Grand Unified Models have only one independent coupling constant, and impose strict algebraic relationships on the couplings of all objects to all fields. They therefore guarantee such a "quantization of charge." It turns out that any theory that produces such a charge spectrum always contains magnetic monopoles. The discovery of a magnetic monopole of the correct mass (as well as other properties predicted by the theory) would therefore be a triumph.*

(b) CP violation and the cosmological proton abundance

Let us now turn to the question of CP violation. When there are three generations, the charge-changing weak interaction involves a 3×3 unitary matrix A, which relates families to generations. If one ignores the third generation, as we did in §9(g), A is 2×2; the phases can then be removed from this unitary matrix by redefining the wave functions of the quarks and leptons, so that A reduces to the orthogonal form given in Eq. (51). In the 3×3 case appropriate to three generations, that can no longer be done. In consequence, the weak interaction contains some complex numbers. A Hamiltonian that contains complex numbers violates time-reversal invariance [see Gottfried (1966), §39]. But as we pointed out in §C.6, the CPT theorem states that the combined operation of space reflection, time reversal, and charge conjugation is always a symmetry of any quantum field theory. Hence a lack of time-reversal invariance implies a violation of CP symmetry. In short, the charge-changing weak interaction of three generations of quarks and leptons accommodates CP violation in a natural manner. On the other

---

* For an (isolated) observation of an event of this type, see Cabrera (1982).

hand, no existing theory provides an explanation as to why CP violation is such a small effect.*

CP violation manifests itself in phenomena which show a visible distinction between a particle and its antiparticle, as Eq. (86) and the related discussion revealed. One may therefore ask whether the existence of CP violation can explain the asymmetry between matter and antimatter observed in the universe. Remarkably enough, Grand Unified Theory, when combined with the Big Bang model of the formation of the universe, provides such a connection.

The argument has the following essential ingredients. First, one assumes that the universe is born as a state having the quantum numbers of the vacuum, and therefore equal numbers of fermions and antifermions. As the universe cools, the mean energy of all constituents drops accordingly. At temperatures above the grand unification mass m_X, all field quanta are involved in fermion–antifermion pair creation and annihilation. Once the temperature drops below m_X, the quanta involved in quark–lepton transitions drop out of the picture since they disappear forever by decay into other particles. (The same happens to the W and Z^0 bosons at a "much" later time when the universe has cooled below 100 GeV.) At first sight one might suppose that the existence of CP violation would assure a fermion–antifermion asymmetry. That is not so. One can show that a system that is always in thermal equilibrium, and which initially has equal numbers of particles and antiparticles, will always retain that symmetry even if CP is violated. That is, if the observed asymmetry is to emerge from an initially symmetric state via CP-violating interactions, the universe must have been out of thermal equilibrium during an epoch when CP-violating processes were significant.

As the universe expands from the Big Bang, its temperature and density fall rapidly, and it will go out of thermal equilibrium if the number of collisions also falls too rapidly. Detailed investigation shows that the interactions of the Grand Unified Theory are such that this occurs only during the relatively brief epoch when the nominal temperature is of order m_X. During that period a computable and very small imbalance between fermions and antifermions is generated. As the system cools further, thermal equilibrium is restored, but the previously produced imbalance cannot be erased by subsequent creation and annihilation processes. If, by definition, we call the more copious species "particles," the ultimate cold state will consist of photons and particles, but no antiparticles, and a very small (and computable) ratio of particles (electrons and nucleons) to photons. This ratio is estimated to be of order 10^{-9}, and is in reasonable agreement with the ratio inferred

---

* If time reversal is not a symmetry, an isolated nondegenerate state of nonzero angular momentum $\mathbf{J}$ can have a permanent electric dipole moment $\mathbf{d}$. This moment must be proportional to the expectation value of $\mathbf{J}$, because that is the only vector that exists in the particle's rest frame. Note that $\mathbf{J}$ is odd under time reversal, as expected. Very sensitive measurements on the neutron (Ramsey, 1982) show that $|d_n| < e \cdot 10^{-10}$ fm. The observation of a nonzero d_n would have significant implications for our understanding of CP violation.

from astronomical observations. Even in our epoch, this ratio is very small because the number of photons in the 3°K black body radiation from the Big Bang is about 10^9 larger than the estimated number of protons (though the energy of the 3° radiation is only one part in 4000 of the total mass of the universe). Hence Grand Unified Theory may be able to dispose of one of the most vexing puzzles of cosmology.

(c) Gravitation

The interaction most familiar to us from everyday life—gravity—has been ignored thus far. Can it be unified with the strong and electroweak interactions into a truly unified field theory? And, for that matter, was it legitimate to ignore gravity in the foregoing considerations? Let us examine this second question first.

At all energies available to us today, the gravitational interaction is negligible compared to the other interactions. But the relative strength of these interactions changes with energy, primarily because the amplitude for scattering between two particles due to gravity grows as a power of the energy, whereas the amplitude due to the other interactions only vary logarithmically. The characteristic energy beyond which gravity must be taken into account is given by the Planck mass, m_P, which is related to Newton's gravitational constant G by

$$m_P = (\hbar c/G)^{\frac{1}{2}} \simeq 10^{19} \text{ GeV}. \tag{93}$$

As this is much larger than m_X, gravitational effects can be ignored in all the phenomena we have discussed thus far.

Nevertheless, it would be esthetically appealing, to say the least, if one could construct a theory that incorporated all the known interactions in a truly integrated fashion. Two paths toward that end are being pursued at present. One conjectures that gravity is not due to an independent quantum field, but is a cooperative effect arising from the Fermi and Bose fields that are already familiar to us. This ingenious approach has not yet been formulated in a really satisfactory fashion.

The second approach to incorporating gravity is, in some ways, more conventional, in that it assumes gravity to be due to a dynamically independent field. It exploits the novel concept of *supersymmetry*, which relates fermions to bosons, in the sense that certain fermions and bosons appear together in multiplets. When Einstein's requirement that the theory be invariant under arbitrary coordinate transformations is imposed on a theory that is supersymmetric, what is called *supergravity* emerges. In supergravity, the quantum of the gravitational field, the spin 2 graviton, appears in a multiplet with fermions and bosons of lower spin. Supergravity is far more constrained than conventional (flat-space) Grand Unified Theory. Indeed, in contrast to the latter, supergravity does not seem to have enough space in its multiplets for the presently observed quarks and leptons. But it does have the intriguing virtue of putting the Einstein gravitational field equations, and the field equations for the strong and electroweak interactions, into one neat package. For that reason alone supergravity merits the intensive investigation to which it is being subjected.

(d) Lepton and quark substructure

It is of course very possible, if not probable, that the quarks and leptons are not the ultimate fermions from which all matter is constructed—that they are themselves composites in the sense that hadrons are composed of quarks. The existence of distinct, yet identical, generations would seem to point in that direction.*

The conjecture that quarks and leptons have substructure, when combined with the present limit of $\sim 10^{-16}$ cm on their dimensions, leads to the following observation. The characteristic excitation energy for a system of such a size is ~ 200 GeV, which is very large compared to the mass splittings between the three known generations. If these separations are themselves caused by internal dynamics, they would probably have to be viewed as the fine structure splitting of the ground state. Should that conjecture be correct, we should think of ourselves as being in the situation of someone attempting to divine the internal structure of the hydrogen atom from the 21 cm radiation emitted in transitions between the hyperfine partners of the hydrogen ground state.

If quarks and leptons are actually composites, many of the speculations that we have discussed would either have to be dropped, or reconsidered.** There would then be new forces which are responsible for binding the new constituents into quarks and leptons. The strong and electroweak forces would be no more fundamental than the chemical or nuclear forces, and the attempt to build a unified field theory from elementary quarks, leptons, photons, gluons, and W's would seem to have been misguided.

Only further experimental research can give us decisive answers to these questions. If the current orthodoxy is correct, and just three generations of

---

* Another hint that we have not seen the ultimate constituents comes from symmetry breaking. As discussed in §10(c), the weak isospin symmetry must be "broken" because W and Z^0 are not degenerate with the photon, and as mentioned there, this breaking can be accomplished by the Higgs mechanism, wherein a spin 0 field acquires a nonzero vacuum expectation value. In the Grand Unified Theories the "grand" symmetry that relates the strong and electroweak interactions must be badly broken, for if it were not, m_X would be comparable to m_W and m_Z. Once again, the only known way of implementing this symmetry breaking is with the Higgs mechanism. At the moment there is no direct experimental evidence for the Higgs particle required by the electroweak symmetry breaking, let alone the "grand" breaking. Furthermore, the Higgs mechanism suffers from a serious inconsistency which has led to the suspicion that there is no fundamental Higgs field, and that the symmetry breaking is due to *new* fermions (i.e., fermions that are neither quarks nor leptons) which form bound states that behave like the Higgs bosons in the naive theory.

** It is also conceivable that the Standard Model will be replaced by a more abstract theory, without conventional fundamental particles, but which, under most circumstances, describes phenomena "as if there were quarks, leptons, etc."

quarks and leptons underlie the structure of all matter, there would be no new phenomena between the energy range set by the electroweak masses m_W and m_Z (or possibly the threshold for t-quark pair production, if that is higher), and the grand unification energy regime at 10^{15} GeV. This "Great Desert" would be devoid of new phenomena, which may not be a happy prospect. On the other hand, if there is quark and lepton substructure, new spectroscopies—the next rungs on the Quantum Ladder—would exist, and could be found.

The discovery of proton decay, or of magnetic monopoles, would be of seminal importance whether or not the lifetime, or the properties of the monopoles, were compatible with Grand Unified Theory. The observation of a process having a characteristic time scale in excess of 10^{31} years would imply the existence of new phenomena in a regime of energies of order 10^{15} GeV! No other natural or man-made phenomena known today can give us access to such energies, or to the interactions that were significant during the epoch when the early universe was that dense and hot.

Appendix I

PARTICLE PROPERTIES

This is a greatly abbreviated version of the authoritative *Tables of Particle Properties* published by the Particle Data Group (1982), except for certain recent data (in particular, for *B*- and *Y*-mesons), in which case references are cited as indicated.

Many of the more accurate results are rounded off, in which case no errors are shown; sometimes errors are indicated in parentheses, as in $3.827(8) = 3.827 \pm 0.008$. In many instances only the most important decay modes are listed.

All masses and widths are in MeV, and all lifetimes in seconds, unless otherwise indicated. Neutrino types are usually not indicated; thus $\mu \to e \nu \bar{\nu}$ means $\mu \to e \nu_\mu \bar{\nu}_e$, etc. Decays such as $\pi^\pm \to \mu \nu$, mean $\pi^+ \to \mu^+ \nu_\mu$ or $\pi^- \to \mu^- \bar{\nu}_\mu$, etc. For many particle–antiparticle pairs, such as $K^\pm$, the decay modes are shown for the positively charged member, as in $K^+ \to \pi^+ \pi^+ \pi^-$, where it is to be understood that the corresponding K^- decay has the charge-conjugate decay products, i.e., $K^- \to \pi^- \pi^- \pi^+$. The notation X stands for anything, as in $D^+ \to K^- X$, where X is $\pi^+ \pi^+$, or $\pi^+ \pi^+ \pi^0$, or

TABLE 1

Field quanta and leptons[a]

Particle	Mass	Mean life	Decay	
			Mode	BR
γ	$<6 \cdot 10^{-22}$	stable		
ν_e	$<5 \cdot 10^{-5}$	stable		
e	0.5110	$>2 \cdot 10^{22}$ yrs		
ν_μ	<0.52	stable		
μ	105.7	2.197×10^{-6}	$e \nu \bar{\nu}$	99%
			$e \nu \bar{\nu} \gamma$	1%
			$e \gamma$	$<2 \times 10^{-10}$
τ	1784(3)	$(3.2 \pm 0.5) \times 10^{-13}$	$\mu \nu \bar{\nu}$	$(19 \pm 1)\%$
			$e \nu \bar{\nu}$	$(16 \pm 1)\%$
			$\pi \nu$	$(11 \pm 2)\%$
			$\rho \nu$	$(22 \pm 4)\%$
$W^\pm$	81 ± 3 GeV		$e \bar{\nu}$	seen
			$\mu \bar{\nu}$	seen
Z^0	96 ± 3 GeV (UA1)	$\Gamma_Z < 8.5$ GeV	$e \bar{e}$	seen
	91 ± 2 GeV (UA2)		$\mu \bar{\mu}$	seen

[a] The τ lifetime is from Jaros (1983); the W mass is from Spiro (1983) and Clark (1983); and the quoted 90% confidence limit on the Z^0 width is from Sadoulet (1983). The UA1 and UA2 values for the Z^0 mass are, respectively, from Sadoulet and Clark.

$\pi^+\pi^+\pi^+\pi^-$, etc. If no branching ratio (BR) is shown, the decay has been observed, but the data do not suffice to determine a numerical value for this ratio.

By "stable hadrons" one means hadrons that only decay via the weak or electromagnetic interaction.

TABLE 2

Stable mesons[a]

Particle	$I^G(J^P)C_n$	Mass	Mean life	Decay	
				Mode	BR
$\pi^\pm$	$1^-(0^-)$	139.6	2.60×10^{-8}	$\mu\nu$	100%
				$e\nu$	1.3×10^{-4}
		$(m_{\pi^\pm} - m_\mu = 33.91)$		$e\nu\pi^0$	1.0×10^{-8}
				$\mu\nu\gamma$	$1.2(3)\times 10^{-4}$
π^0	$1^-(0^-)+$	135.0	$0.83(6) \times 10^{-16}$	$\gamma\gamma$	98.8%
				$\gamma e\bar{e}$	1.2%
		$(m_{\pi^\pm} - m_{\pi^0} = 4.60)$		$e\bar{e}e\bar{e}$	3×10^{-5}
η	$0^+(0^-)+$	548.8(6)	$\Gamma = 0.8(1)$ keV	$\gamma\gamma$	39%
				$3\pi^0$	32%
				$\pi^+\pi^-\pi^0$	24%
				$\pi^+\pi^-\gamma$	4.9(1)%
				$e^+e^-\gamma$	0.5(1)%
$K^\pm$	$\frac{1}{2}(0^-)$	493.7	1.24×10^{-8}	$\mu\nu$	64%
				$\pi^+\pi^0$	21%
				$\pi^+\pi^+\pi^-$	5.6%
				$\pi^+\pi^0\pi^0$	1.7%
				$\pi^0\mu\nu$	3.2%
				$\pi^0 e\nu$	4.8%
				$\pi^+ e\bar{e}$	3×10^{-7}
				$\pi^- e^+e^+$	$<10^{-8}$
				$\pi^+\gamma\gamma$	$<4 \times 10^{-5}$
				$\pi^+\gamma$	$<4 \times 10^{-6}$
				$\pi^- e^+\mu^+$	$<10^{-8}$
				$\pi^+ e^-\mu^+$	$<5 \times 10^{-9}$
				$\mu^+ e^+ e^-\nu$	$11(3) \times 10^{-7}$
				$\mu^- e^+ e^+\nu$	$<2 \times 10^{-8}$
$K^0, \bar{K}^0$	$\frac{1}{2}(0^-)$	497.7(1)		50% K_S^0, 50% K_L^0	
K_S^0	$\frac{1}{2}(0^-)$		$0.892(2) \times 10^{-10}$	$\pi^+\pi^-$	69%
				$\pi^0\pi^0$	31%
				$\pi^+\pi^-\gamma$	2×10^{-3}
				$\mu^+\mu^-$	$<3 \times 10^{-7}$
				e^+e^-	$<3 \times 10^{-4}$

[a] Here *I, G, J*, and *P* are, respectively, the isospin, *G*-parity, spin, and spatial parity, where *G* is defined in §III.A.9 (Vol. II). C_n is the signature under charge conjugation of the neutral member of the multiplet (see §II.C.7 and §III.A.9, Vol. II). Underlined quantum numbers are those favored by the quark model, but not yet determined experimentally. The data on the *B*-mass and decay information come from the CLEO Collaboration: S. Behrends et al., *Phys. Rev. Lett.* **50**, 881 (1983), and K. Chadwick et al., *Phys. Rev.* **D27**, 475 (1983). The lifetimes of *B*-, *D*-, and *F*-mesons are from the compilation of Reay (1983); the lifetime quoted for *B* is actually that for hadrons containing *b*-quarks, i.e., either mesons or baryons (see Lockyer, 1983). The *F*-mass is from A. Chen et al., *Phys. Rev. Lett.* **51**, 634 (1983).

TABLE 2 (*continued*)

Particle	$I^G(J^P)C_n$	Mass	Mean life	Decay	
				Mode	BR
K_L^0	$\frac{1}{2}(0^-)$		$5.18(4) \times 10^{-8}$	$\pi^0\pi^0\pi^0$	$(22 \pm 1)\%$
				$\pi^+\pi^-\pi^0$	12%
		$m(K_L) - m(K_S) = 3.52(1) \times 10^{-6}$ eV		$\pi\mu\nu$	27%
				$\pi e\nu$	39%
				$\pi^+\pi^-$	$0.203(5)\%$
				$\pi^0\pi^0$	$0.09(2)\%$
				$\mu^+\mu^-$	$(9 \pm 2) \times 10^{-9}$
				e^+e^-	$<2 \times 10^{-7}$
				$e^+e^-\gamma$	$(2 \pm 1) \times 10^{-5}$
$D^\pm$	$\frac{1}{2}(0^-)$	1869.4(6)	$(8 \pm 1) \times 10^{-13}$	eX	$(19 \pm 4)\%$
				K^-X	$(16 \pm 4)\%$
				K^+X	$(6 \pm 3)\%$
				$K^0X + \bar{K}^0X$	$(48 \pm 15)\%$
				$\underline{K}^-\pi^+\pi^+$	$(5 \pm 1)\%$
				$\bar{K}^0\pi^+$	$1.8(5)\%$
				$\bar{K}^0\pi^+\pi^0$	$(13 \pm 8)\%$
				$\bar{K}^0\pi^+\pi^+\pi^-$	$(8 \pm 4)\%$
				$\bar{K}^0K^+$	$0.5(3)\%$
$D^0, \bar{D}^0$	$\frac{1}{2}(0^-)$	1864.7(6)	$3.8(4) \times 10^{-13}$	eX	$<6\%$
				K^-X	$(44 \pm 10)\%$
				$K^0X + \bar{K}^0X$	$(33 \pm 10)\%$
				K^+X	$(8 \pm 3)\%$
				$\underline{K}^-\pi^+\pi^0$	$(9 \pm 3)\%$
				$\bar{K}^0\pi^0$	$(2 \pm 1)\%$
				$K^-\rho^+$	$(7 \pm 3)\%$
				$\pi^+\pi^-$	$(8 \pm 4) \times 10^{-4}$
				K^+K^-	$(3 \pm 1) \times 10^{-3}$
$F^\pm$	$0(0^-)$	1970 ± 7	$(3 \pm 1) \times 10^{-13}$	$\eta\pi, \eta\pi\pi\pi, \rho^+\phi$	
$B^0, \bar{B}^0$	$\frac{1}{2}(0^-)$	5274 ± 4	$\sim 2 \times 10^{-12}$	charmed states dominant	
$B^\pm$	$\frac{1}{2}(0^-)$	5271 ± 4			

TABLE 3

Stable baryons[a]

Particle	$I(J^P)$	Mass	Mean life	Decay Mode	Decay BR
p	$\frac{1}{2}(\frac{1}{2}^+)$	938.3	$>2 \times 10^{31}$ yrs		
n	$\frac{1}{2}(\frac{1}{2}^+)$	939.6	925 ± 11	$pe\nu$	100%
	$(m_p - m_n = 1.293)$				
Λ	$0(\frac{1}{2}^+)$	1115.6	$2.63(2) \times 10^{-10}$	$p\pi^-$	64%
				$n\pi^0$	36%
				$pe^-\nu$	$8.4(2) \times 10^{-4}$
				$p\mu^-\nu$	$1.6(4) \times 10^{-4}$
Σ^+	$1(\frac{1}{2}^+)$	1189.4	0.80×10^{-10}	$p\pi^0$	52%
				$n\pi^+$	48%
				$\Lambda e^+\nu$	$2.0(5) \times 10^{-5}$
Σ^0	$1(\frac{1}{2}^+)$	1192.5(1)	$(6 \pm 1) \times 10^{-20}$	$\Lambda\gamma$	100%
Σ^-	$1(\frac{1}{2}^+)$	1197.3(1)	$1.48(1) \times 10^{-10}$	$n\pi^-$	$\approx$100%
				$ne^-\nu$	1.1×10^{-3}
				$n\mu^-\nu$	5×10^{-4}
				$\Lambda e^-\nu$	6×10^{-5}
Ξ^0	$\frac{1}{2}(\frac{1}{2}^+)$	1314.9(6)	$2.9(1) \times 10^{-10}$	$\Lambda\pi^0$	$\approx$100%
Ξ^-	$\frac{1}{2}(\frac{1}{2}^+)$	1321.3(1)	1.6×10^{-10}	$\Lambda\pi^-$	$\approx$100%
				$\Lambda e^-\nu$	$(3 \pm 1) \times 10^{-4}$
				$\Lambda\mu^-\nu$	$(4 \pm 4) \times 10^{-4}$
Ω^-	$0(\frac{3}{2}^+)$	1672.5(3)	$0.82(3) \times 10^{-10}$	ΛK^-	$(69 \pm 1)\%$
				$\Xi^0\pi^-$	$(23 \pm 1)\%$
				$\Xi^-\pi^0$	$(8 \pm 1)\%$
Λ_c^+	$0(\frac{1}{2}^+)$	2282 ± 3	$(2.2^{+0.7}_{-0.4}) \times 10^{-13}$	$pK^-\pi^+$	$(2 \pm 1)\%$
				$p\bar{K}^0$	$(1 \pm 1)\%$
				$\Lambda\pi^+, \Sigma^0\pi^+$	
				$\Delta^{++}K^-$	

[a] For Ξ, P has not been measured, while for Ω^- and Λ_c^+ neither J nor P is measured. Quantum numbers shown are those favored by quark model. The limit on the free proton lifetime is a 90% confidence level measurement on the mode $p \to e^+\pi^0$ by R. M. Bionta (1983). For Λ_c^+ lifetime, see Reay (1983).

TABLE 4
Mesons without strangeness or charm, or heavy quark constituents

Particle	$I^G(J^P)C_n$	Mass	Width	Mode	BR (%)
$\pi^\pm$	$1^-(0^-)+$	139.6	0.0	See Table 2	
π^0		135.0	8.0(6) eV		
η	$0^+(0^-)+$	549	0.8(1) keV		
ρ	$1^+(1^-)-$	769 ± 3	154 ± 5	$\pi\pi$	≈ 100
				$\mu\bar{\mu}$	0.007(1)
				$e\bar{e}$	0.0043(5)
ω	$0^-(1^-)-$	783	9.9(3)	$\pi^+\pi^-\pi^0$	90
				$\pi^0\gamma$	9
				$\pi^+\pi^-$	1
				$e\bar{e}$	0.0072(7)
η'	$0^+(0^-)+$	958	0.3(1)	$\eta\pi\pi$	65 ± 2
				$\rho^0\gamma$	30 ± 2
				$\omega\gamma$	2.8(5)
				$\gamma\gamma$	1.9(2)
S^*	$0^+(0^+)+$	975 ± 4	33 ± 6	$\pi\pi$	78 ± 3
				$K\bar{K}$	22 ± 3
δ	$1^-(0^+)+$	983 ± 2	54 ± 7	$\eta\pi$ & $K\bar{K}$	
ϕ	$0^-(1^-)-$	1020	4.2(1)	K^+K^-	49 ± 1
				$K_L K_S$	35 ± 1
				$\pi^+\pi^-\pi^0$	15 ± 1
				$\eta\gamma$	1.5(2)
				$\pi^0\gamma$	0.14(5)
				$e\bar{e}$	0.031(1)
				$\mu\bar{\mu}$	0.025(3)
f	$0^+(2^+)+$	1273 ± 5	179 ± 20	$\pi\pi$	83 ± 2
				$2\pi^+2\pi^-$	2.8(4)
				$K\bar{K}$	2.9(2)
$A1$	$1^-(1^+)+$	1275 ± 30	315 ± 45	$\rho\pi$ & $\pi(\pi\pi)_{s\text{-wave}}$	
ε		~ 1300	200–600	$\pi\pi$	~ 90
				$K\bar{K}$	~ 10
$A2$	$1^-(2^+)+$	1318 ± 5	110 ± 5	$\rho\pi$	70 ± 2
				$\eta\pi$	15 ± 1
				$\omega\pi\pi$	11 ± 3
				$K\bar{K}$	4.8(5)
f'	$0^+(2^+)+$	1520 ± 10	75 ± 10	$K\bar{K}$	dominant
ρ'	$1^+(1^-)-$	1600 ± 20	300 ± 100	4π	dominant
$\omega(1670)$	$0^-(3^-)-$	1688 ± 5	166 ± 15	3π & 5π	
$A3$	$1^-(2^-)+$	1680 ± 30	250 ± 50	$f\pi$	55 ± 5
				$\rho\pi$	36 ± 6
$\phi(1680)$	$0^-(1^-)-$	1684 ± 15	126 ± 22	$K^*\bar{K} + \bar{K}^*K$	dominant
g	$1^+(3^-)-$	1691 ± 5	200 ± 20	2π	24 ± 1
				4π	71 ± 2
				$K\bar{K}\pi$	4 ± 1
				$K\bar{K}$	2
h	$0^+(4^+)+$	2040 ± 20	150 ± 50	$\pi\pi$ & $K\bar{K}$	

TABLE 5
Mesons containing one s, c, or b quark

Particle	$I^G(J^P)$	Mass	Width	Decay Mode	BR(%)
$K^\pm$	$\frac{1}{2}(0^-)$	493.7		See Table 2	
K^0		497.7			
K^*	$\frac{1}{2}(1^-)$	892[a]	51 ± 1	$K\pi$	≈ 100
				$K\gamma$	0.15(7)
$Q_1(1280)$	$\frac{1}{2}(1^+)$	1270 ± 10	90 ± 20	$K\rho$	42 ± 6
				$K\pi$	28 ± 4
				$K^*\pi$	16 ± 5
				$K\omega$	11 ± 2
$\kappa(1350)$	$\frac{1}{2}(0^+)$	~ 1350	~ 250	$K\pi$	
$K^*(1430)$	$\frac{1}{2}(2^+)$	1434 ± 5	100 ± 10	$K\pi$	45 ± 2
				$K^*\pi$ & $K^*\pi\pi$	38 ± 3
				$K\rho$	9 ± 1
				$K\omega$	4 ± 2
$K^*(1780)$	$\frac{1}{2}(3^-)$	1775 ± 10	140 ± 20	$K\pi\pi$[b]	dominant
				$K\pi$	17 ± 5
D, F			See Table 2		
D^{*+}	$\frac{1}{2}(1^-)$	2010	<2	$D^0\pi^+$	64 ± 11
				$D^+\pi^0$	28 ± 9
				$D^+\gamma$	8 ± 7
D^{*0}	$\frac{1}{2}(1^-)$	2007	<5	$D^0\pi^0$	55 ± 15
				$D^0\gamma$	45 ± 15
B			See Table 2		

[a] Mass is for $K^{*\pm}$; $m(K^{*0}) - m(K^{*\pm}) = 7 \pm 1$.
[b] $K\pi\pi$ includes $K\rho$ and $K^*\pi$.

TABLE 6

Mesons having the composition $Q\overline{Q}$[a]

Here Q is a heavy quark—either c or b. Instead of specifying the measured quantity J^P, we give the quark model quantum numbers $^{2S+1}L_J$, where $S = 0$ or 1 is the total intrinsic quark spin, L is their relative orbital angular momentum ($L = 0, 1, 2, \ldots$ are designated by S, P, D, etc.), and the parity is given by $(-1)^{L+1}$. When underlined, the quantum number has not been confirmed by experiment. All these states have $I = 0$.

Particle	$G(^{2S+1}L_J)C$	Mass	Width	Decay	
				Mode	BR (%)
η_c	$+(^1S_0)+$	2981 ± 6	<20	$\eta\pi^+\pi^-$, 4π, 2π, $2K$, $p\bar{p}$	
η_c'	$+(^1S_0)+$	3590			
J/ψ	$-(^3S_1)-$	3097	(63 ± 9) keV	$e\bar{e}$	7 ± 1
				$\mu\bar{\mu}$	7 ± 1
				5, 7, or 9 π's	8 ± 1
				$K^+K^-\pi^+\pi^-\pi^0$	1.2(3)
				4 or 6 π's	0.8(3)
				$p\bar{p}\pi^-\pi^+$	0.5(1)
				$\Xi\bar{\Xi}$	0.3(1)
				$\rho\pi$	1.2(1)
				$\omega 2\pi^+2\pi^-$	0.9(3)
				$\omega\pi\pi$	0.7(2)
				ρA_2	0.8(5)
				$K^*(892)\bar{K}^*(1430) \pm$ c.c.	0.7(3)
				ωf	0.2(1)
				$\omega p\bar{p}$	0.2
				and many others	

		Mass	Width	Decay modes	
$\psi(3685)$ ($\equiv \psi'$)	$-(^3S_1)-$	3686	(215 ± 40) keV	$e\bar{e}$	0.9(1)
				$\mu\bar{\mu}$	0.8(2)
				hadrons $+ \gamma$'s	98
				$\gamma\chi(3415)$	8 ± 1
				$\gamma\chi(3510)$	8 ± 1
				$\gamma\chi(3555)$	7 ± 1
				$\gamma\eta_c$	0.4(3)
				$\gamma\eta_c'$	seen
				$\psi\pi^+\pi^-$	33 ± 2
				$\psi\pi^0\pi^0$	17 ± 2
				$2(\pi^+\pi^-)\pi^0$	0.4(2)
				$\pi^+\pi^-K^+K^-$	0.2
				$p\bar{p}\pi^+\pi^-$	0.08(2)
				and many others	
$\psi(3770)$	$-(^3\underline{D_1})-$	3770 ± 3	25 ± 3	$e\bar{e}$	0.001
				$D\bar{D}$	dominant
$\psi(4030)$	$-(^3S_1)-$	4030 ± 5	50 ± 10	$e\bar{e}$	0.001
				hadrons	dominant
$\psi(4160)$	$-(^3S_1)-$	4160 ± 20	80 ± 20	$e\bar{e}$	0.001
				hadrons	dominant
$\psi(4415)$	$-(^3S_1)-$	4415 ± 6	40 ± 20	$e\bar{e}$	0.001
				hadrons	dominant
$\chi(3415)$	$+(^3P_0)+$	3415 ± 1		$2(\pi^+\pi^-)$	4 ± 1
				$\pi^+\pi^-K^+K^-$	3 ± 1
				$3(\pi^+\pi^-)$	2 ± 1
				$\pi^+\pi^-$	0.9(2)
				$\gamma\psi$	0.8(2)
				K^+K^-	0.8(2)
$\chi(3510)$	$+(^3P_1)+$	3510 ± 1		$\gamma\psi$	28 ± 3
				$3(\pi^+\pi^-)$	2.4(9)
				$2(\pi^+\pi^-)$	1.8(5)
				$\pi^+\pi^-K^+K^-$	1.0(4)

[a] Data on Υ- and χ_b-states not listed in Particle Data Group are from A. S. Artamanov et al., *Novosibirsk Report No. 82–94*, Academy of Sciences of the USSR; CLEO Collaboration, D. Andrews et al., *Phys. Rev. Lett.* **50**, 807 (1983); P. Franzini and J. Lee-Franzini, *Ann. Rev. Nucl. Part. Sci.* **33**, 1 (1983); K. Berkelman, *Physics Reports C* (1983), in press; Gottfried (1983); and Tuts (1983). The masses of $\chi_b(1P)$ and $\chi_b(2P)$ given are the center-of-gravity of the multiplet. The photon spectra observed in $\Upsilon(3S) \to \gamma X$ and $\Upsilon(2S) \to \gamma X$ show that these are multiplets, but the splittings are not fully established as yet (see Tuts, 1983).

TABLE 6 (continued)

Particle	$G(^{2S+1}L_J)C$	Mass	Width	Decay Mode	BR (%)
$\chi(3556)$	$+(^3P_2)+$	3556 ± 1		$\gamma\psi$	16 ± 2
				$2(\pi^+\pi^-)$	$2.3(5)$
				$3(\pi^+\pi^-)$	$1.2(8)$
				$\pi^+\pi^-K^+K^-$	$2.0(5)$
$Y(1S) \equiv Y$	$\underline{-(^3S_1)-}$	9460	48 ± 8 keV	$e\bar{e}$	5.1 ± 3.0
				$\mu\bar{\mu}$	3.0 ± 0.3
				$\tau\bar{\tau}$	3.4 ± 0.7
$Y(2S) \equiv Y'$	$\underline{-(^3S_1)-}$	10020 ± 10	27 ± 23 keV	$e\bar{e}$	1.9 ± 1.8
				$\mu\bar{\mu}$	1.9 ± 1.4
				$Y(1S)X$	30 ± 6
				$\gamma\chi_b(1P)$	15.5 ± 2.5
				$\pi\pi Y(1S)$	30 ± 4
				$\gamma\gamma Y(1S)$	3.6 ± 0.9
$Y(3S) \equiv Y''$	$\underline{-(^3S_1)-}$	10351 ± 10	13 ± 7 keV	$e\bar{e}$	3 ± 2
				$\mu\bar{\mu}$	2.9 ± 1.0
				$\gamma\chi_b(1P)$	36 ± 3
				$\gamma\chi_b(2P)$	7.0 ± 1.4
				$\pi\pi Y(1S)$	4.6 ± 3.0
				$\pi\pi Y(2S)$	3.6 ± 1.2
				$\gamma\gamma Y(1S)$	5.9 ± 2.1
				$\gamma\gamma Y(2S)$	
$Y(4S)$	$\underline{-(^3S_1)}$	10569 ± 10 $(\Gamma_{e\bar{e}} = 0.27 \pm 0.04$ keV$)$	14 ± 5 MeV	$e\bar{e}$	
				$B\bar{B}$	
$\chi_b(1P) \equiv \chi_b$	$\underline{+(^3P)+}$	9900 ± 3		$\gamma Y(1S)$	
$\chi_b(2P) \equiv \chi_b'$	$\underline{+(^3P)+}$	10256 ± 5		$\gamma Y(1S)$	
				$\gamma Y(2S)$	

TABLE 7

Nucleon levels $(I = \frac{1}{2})$

The higher-lying levels are observed as resonances in the phase of πN scattering amplitudes (see §III.A, Vol. II). If the resonance is broad, this often does not allow one to extract a one standard deviation error for the mass and width, even though the existence of the resonance, and its spin-parity, can be established. In this and the following baryon tables we use the notation defined in Table E.1, and therefore do not specify I or S.

Particle	J^P	Mass	Width	Decay Mode	BR (%)
p		938.3		See Table 3	
n	$\frac{1}{2}^+$	939.6			
$N(1440)$	$\frac{1}{2}^+$	1440 ± 40	120–350	$N\pi$	50–70
				$N\eta$	8–18
				$N\pi\pi$	~30
$N(1520)$	$\frac{3}{2}^-$	1520 ± 10	125 ± 15	$N\pi$	50–60
				$N\pi\pi$	35–50
				$N\rho$	15–25
				$\Delta\pi$	15–25
$N(1535)$	$\frac{1}{2}^-$	1540 ± 20	100–250	$N\pi$	35–50
				$N\eta$	40–65
				$N\pi\pi$	~5
$N(1650)$	$\frac{1}{2}^-$	1650 ± 30	100–200	$N\pi$	55–65
				ΛK	5–10
				ΣK	3–10
				$N\rho$	~20
				$\Delta\pi$	4–15
$N(1675)$	$\frac{5}{2}^-$	1675 ± 15	150 ± 30	$N\pi$	30–40
				$\Delta\pi$	50–65
$N(1680)$	$\frac{5}{2}^+$	1680 ± 10	125 ± 15	$N\pi$	55–65
				$\Delta\pi$	~12
				$N\rho$	~10
				$N\varepsilon$	~20
$N(1700)$	$\frac{3}{2}^-$	1700 ± 30	100 ± 30	$N\pi$	8–12
				$N\eta$	~4
				$\Delta\pi$	15–40
				$N\rho$	~5
$N(1710)$	$\frac{1}{2}^+$	1710 ± 30	110 ± 20	$N\pi$	10–20
				$N\eta$	5–35
				ΛK	5–15
				ΣK	2–10
				$\Delta\pi, N\rho, N\varepsilon$	>50
$N(1720)$	$\frac{3}{2}^+$	1690–1800	125–250	$N\pi$	10–20
				$N\eta$	3–6
				$\Lambda K, \Sigma K$	4–17
				$\Delta\pi, N\rho, N\varepsilon$	~70
$N(1990)$	$\frac{7}{2}^+$	1950–2050	120–400	$N\pi$	~5
				$N\eta$	~3
				$\Lambda K, \Sigma K$	
$N(2200)$	$\frac{5}{2}^-$	1900–2230	150–400	$N\pi$	~8
				$N\eta, \Lambda K$	
$N(2220)$	$\frac{9}{2}^+$	2150–2300	300–500	$N\pi$	~18
$N(2250)$	$\frac{9}{2}^-$	2130–2270	200–500	$N\pi$	~10
				$N\eta$	~2
$N(2600)$	$\frac{11}{2}^-$	2580–2700	>300	$N\pi$	~5

TABLE 8
Δ levels ($I = \frac{3}{2}$)

Particle	J^P	Mass	Width	Decay	
				Mode	BR(%)
$\Delta(1232)$	$\frac{3}{2}^+$	1232 ± 2	110–120	$N\pi$	99.4
				$N\gamma$	0.6
$\Delta(1600)$	$\frac{3}{2}^+$	1500–1900	150–350	$N\pi$	15–25
				$N\pi\pi$, $\Delta\pi$, $N\rho$	~80
$\Delta(1620)$	$\frac{1}{2}^-$	1600–1650	120–160	$N\pi$	25–35
				$N\pi\pi$, $\Delta\pi$, $N\rho$	~70
$\Delta(1700)$	$\frac{3}{2}^-$	1630–1740	190–300	$N\pi$	10–20
				$N\pi\pi$, $\Delta\pi$, $N\rho$	~85
$\Delta(1900)$	$\frac{1}{2}^-$	1850–2000	130–300	$N\pi$	6–12
				ΣK	~10
$\Delta(1905)$	$\frac{5}{2}^+$	1890–1920	250–400	$N\pi$	8–15
				$N\pi\pi$, $\Delta\pi$, $N\rho$	~80
$\Delta(1910)$	$\frac{1}{2}^+$	1850–1950	200–330	$N\pi$	23
				ΣK	2–20
				$N\pi\pi$, $N\rho$	>40
$\Delta(1920)$	$\frac{3}{2}^+$	1860–2160	190–300	$N\pi$	~16
				ΣK	~5
$\Delta(1930)$	$\frac{5}{2}^-$	1890–1960	150–350	$N\pi$	4–14
				ΣK	<10
$\Delta(1950)$	$\frac{7}{2}^+$	1910–1960	200–340	$N\pi$	35–45
				$N\pi\pi$, $\Delta\pi$, $N\rho$	~60
$\Delta(2420)$	$\frac{11}{2}^+$	2380–2450	300–500	$N\pi$	5–15

TABLE 9
Strange baryon levels

Particle	J^P	Mass	Width	Decay	
				Mode	BR(%)
Λ	$\frac{1}{2}^+$	1116		See Table 3	
$\Lambda(1405)$	$\frac{1}{2}^-$	1405 ± 5	40 ± 10	$\Sigma\pi$	≈100
$\Lambda(1520)$	$\frac{3}{2}^-$	1519 ± 1	16 ± 1	$N\bar{K}$	45 ± 1
				$\Sigma\pi$	42 ± 1
				$\Lambda\pi\pi$	10 ± 1
$\Lambda(1600)$	$\frac{1}{2}^+$	1560–1700	50–250	$N\bar{K}$	15–30
				$\Sigma\pi$	10–60
$\Lambda(1670)$	$\frac{1}{2}^-$	1670 ± 10	25–50	$N\bar{K}$	15–25
				$\Sigma\pi$	20–60
				$\Lambda\eta$	15–35
$\Lambda(1690)$	$\frac{3}{2}^-$	1690 ± 5	60 ± 10	$N\bar{K}$	20–30
				$\Sigma\pi$	20–40
				$\Lambda\pi\pi$, $\Sigma\pi\pi$	~45
$\Lambda(1800)$	$\frac{1}{2}^-$	1720–1850	200–400	$N\bar{K}$	25–40
				$\Sigma\pi$, $\Sigma(1385)\pi$, $N\bar{K}^*(892)$	
$\Lambda(1820)$	$\frac{5}{2}^+$	1815–1825	80 ± 10	$N\bar{K}$	~60
				$\Sigma\pi$, $\Sigma(1385)\pi$	
$\Lambda(1830)$	$\frac{5}{2}^-$	1810–1830	60–110	$\Sigma\pi$	35–75
				$N\bar{K}$, $\Sigma(1385)\pi$	>15

TABLE 9 (*continued*)

Particle	J^P	Mass	Width	Decay Mode	BR (%)
$\Lambda(1890)$	$\frac{3}{2}^+$	1850–1910	60–200	$N\bar{K}$	20–35
				$\Sigma\pi$	3–10
$\Lambda(2100)$	$\frac{7}{2}^-$	2090–2110	100–250	$N\bar{K}$	25–35
				$N\bar{K}^*(892)$	10–20
$\Lambda(2110)$	$\frac{5}{2}^+$	2090–2140	150–250	$N\bar{K}$	5–25
				$\Sigma\pi$	10–40
				$N\bar{K}^*(892)$	10–60
$\Lambda(2350)$	$\frac{9}{2}^+$	2340–2370	100–250	$N\bar{K}$	~12
				$\Sigma\pi$	~10
Σ	$\frac{1}{2}^+$	1189		See Table 3	
$\Sigma(1385)$	$\frac{3}{2}^+$	1382	35 ± 1	$\Lambda\pi$	88 ± 2
				$\Sigma\pi$	12 ± 2
$\Sigma(1660)$	$\frac{1}{2}^+$	1630–1690	40–200	$N\bar{K}$	10–30
				$\Lambda\pi,\ \Sigma\pi$	
$\Sigma(1670)$	$\frac{3}{2}^-$	1665–1685	40–80	$\Sigma\pi$	30–60
				$N\bar{K}$	~10
				$\Lambda\pi$	5–15
$\Sigma(1750)$	$\frac{1}{2}^-$	1730–1800	60–160	$N\bar{K}$	10–40
				$\Sigma\eta$	15–55
$\Sigma(1775)$	$\frac{5}{2}^-$	1770–1780	105–135	$N\bar{K}$	~40
				$\Lambda\pi$	14–20
				$\Sigma(1385)\pi$	~10
				$\Lambda(1520)\pi$	~20
$\Sigma(1915)$	$\frac{5}{2}^+$	1900–1935	80–160	$N\bar{K}$	5–15
				$\Lambda\pi,\ \Sigma\pi,\ \Sigma(1385)\pi$	
$\Sigma(1940)$	$\frac{3}{2}^-$	1900–1950	150–300	$N\bar{K}$	<20
				$\Lambda\pi,\ \Sigma\pi,\ \Sigma(1385)\pi,$	
				$\Lambda(1520)\pi,\ \Delta(1232)\bar{K},$	
				$N\bar{K}^*(892)$	
$\Sigma(2030)$	$\frac{7}{2}^+$	2025–2040	150–200	$N\bar{K}$	~20
				$\Lambda\pi$	~20
				$\Sigma\pi$	5–10
				$\Sigma(1385)\pi$	5–15
				$\Lambda(1520)\pi$	10–20
				$\Delta(1232)\bar{K}$	10–20
Ξ	$\frac{1}{2}^+$	1315		See Table 3	
$\Xi(1530)$	$\frac{3}{2}^+$	1532	9.1(5)	$\Xi\pi$	≈100
$\Xi(1820)$	$\frac{3}{2}$	1823 ± 6	20^{+15}_{-10}	$\Lambda\bar{K}$	~45
				$\Sigma\bar{K}$	~10
				$\Xi(1530)\pi$	~45

BIBLIOGRAPHY

Ankenbrandt, C. et al. (1983), *Phys. Rev. Lett.* **51**, 863.

Banner, M. et al. (1982), *Phys. Lett.* **118B**, 203.

Bionta, R. M. et al. (1983), *Phys. Rev. Lett.* **51**, 27.

Cabenda, R. (1982), Cornell University Thesis (unpublished).

Cabrera, B. (1982), *Phys. Rev. Lett.* **48**, 1378.

Carrigan, R. A. and W. P. Trower (1982), *Scientific American* **246**, April, p. 106.

Clark, A. (1983), Proc. 1983 Intl. Symp. Photon & Lepton Interactions at High Energy, Cornell University, Ithaca, New York; D. G. Cassel and D. L. Kreinick, editors (to be published); referred to henceforth by Cornell Photon-Lepton Symposium.

Close, F. E. (1979), *Quarks and Partons*, Academic Press, London.

Crawford, F. et al. (1957), *Phys. Rev.* **108**, 1102.

Dirac, P. A. M. (1958), *Quantum Mechanics*, 4th ed., Oxford University Press, London.

Dydak, F. (1978), *Acta Physica Austriaca*, Suppl. XIX, 463.

Georgi, H. (1981), *Scientific American* **244**, April, p. 48.

Georgi, H. and S. L. Glashow (1980), *Physics Today* **33**, September, p. 30.

Gjesdal, S. et al. (1974), *Phys. Lett.* **52B**, 113.

Goldhaber, M., P. Langacker, and R. Slansky (1980), *Science* **210**, 851.

Gottfried, K. (1966), *Quantum Mechanics*, Benjamin, New York.

Gottfried, K. (1970), in *Elementary Particle Physics and Scattering Theory*, Vol. 2, editors M. Chrétien and S. S. Schweber, Gordon & Breach, New York; pp. 125–200.

Gottfried, K. (1983), Proc. Intl. Europhysics Conf. High Energy Phys., Brighton, U.K., to be published by Rutherford-Appleton Laboratory.

Harari, H. (1982), *Scientific American* **248**, April, p. 48.

Hertzog, D. W. et al. (1983), *Phys. Rev. Lett.* **51**, 1131.

Jaros, J. A. et al. (1983), *Phys. Rev. Lett.* **51**, 955.

Kim, J. E., P. Langacker, M. Levine, and H. H. Williams (1981), *Rev. Mod. Phys.* **53**, 211.

LaRue, G. S., J. D. Phillips, and W. M. Fairbank (1981), *Phys. Rev. Lett.* **46**, 967.

Lee, T. D. (1981), *Particle Physics and Introduction to Field Theory*, Harwood, New York.

Lee, T. D. and C. S. Wu (1966), *Annual Reviews of Nuclear Science* **16**, 511–590.

Lockyer, N. et al. (1980), *Phys. Rev. Lett.* **45**, 1821.

Marciano, W. J. (1983), Cornell Photon-Lepton Symposium, *loc. cit.*

MARK J Collaboration (1980), *Phys. Reports* **63**, No. 7.

Particle Data Group (1982), *Phys. Lett.* **111B**, April.

Perkins, D. H. (1982), *Introduction to High Energy Physics*, 2nd ed., Addison-Wesley Reading, Mass.

Ramsey, N. F. (1982), *Reports on Progress in Physics* **45,** 95.

Reay, N. (1983), Cornell Photon-Lepton Symposium, *loc. cit.*

Sadoulet, B. (1983), Cornell Photon-Lepton Symposium, *loc. cit.*

Salam, A. (1980), *Science* **210,** 723.

Schramm, D. N. (1983), *Physics Today* **36,** April, p. 27.

Schwinberg, P. B., R. S. Van Dyck, Jr., and H. G. Dehmelt (1981), *Phys. Rev. Lett.* **47,** 1679.

SLAC-MIT Collaboration (1968), presented at the 14th International Conference on High-Energy Physics, Vienna; Proceedings edited by J. Prentki and J. Steinberger, *CERN,* 1968.

Spiro, W. M. (1983), Cornell Photon-Lepton Symposium, *loc. cit.*

TASSO Collaboration, R. Brandelik et al. (1980), *Phys. Lett.* **94B,** 259.

Turner, M. and D. Schramm (1979), *Physics Today* **32,** September, p 43.

Tuts, P. M. (1983), Cornell Photon-Lepton Symposium, *loc. cit.*

UA1 Collaboration (1983a), G. Arnison et al., *Phys. Lett.* **122B,** 103.

UA1 Collaboration (1983b), G. Arnison et al., *Phys. Lett.* **126B,** 398.

UA2 Collaboration (1983), M. Banner et al., *Phys. Lett.* **122B,** 476.

Weinberg, S. (1977), *The First Three Minutes*, Basic Books, New York.

Weinberg, S. (1980), *Scientific American* **244,** June, p. 64.

Weisskopf, V. F (1981), *Physics Today* **34,** November, p. 69.

Weisskopf, V. F. (1983), *American Scientist* **71,** 473.

Werner, S. A. (1980), *Physics Today* **33,** December, p. 24.

Wick, G. C. (1966), in *Preludes in Theoretical Physics*, editors A. DeShalit, H. Feshbach, and L. Van Hove, North-Holland, Amsterdam.

Wilczek, F. (1980), *Scientific American* **243,** December, p. 82.

INDEX